Lyudmila Khrabrova

Characterization of genetic horse breeding resources in Russia

Lyudmila Khrabrova

Characterization of genetic horse breeding resources in Russia

LAP LAMBERT Academic Publishing

Impressum / Imprint
Bibliografische Information der Deutschen Nationalbibliothek: Die Deutsche Nationalbibliothek verzeichnet diese Publikation in der Deutschen Nationalbibliografie; detaillierte bibliografische Daten sind im Internet über http://dnb.d-nb.de abrufbar.
Alle in diesem Buch genannten Marken und Produktnamen unterliegen warenzeichen-, marken- oder patentrechtlichem Schutz bzw. sind Warenzeichen oder eingetragene Warenzeichen der jeweiligen Inhaber. Die Wiedergabe von Marken, Produktnamen, Gebrauchsnamen, Handelsnamen, Warenbezeichnungen u.s.w. in diesem Werk berechtigt auch ohne besondere Kennzeichnung nicht zu der Annahme, dass solche Namen im Sinne der Warenzeichen- und Markenschutzgesetzgebung als frei zu betrachten wären und daher von jedermann benutzt werden dürften.

Bibliographic information published by the Deutsche Nationalbibliothek: The Deutsche Nationalbibliothek lists this publication in the Deutsche Nationalbibliografie; detailed bibliographic data are available in the Internet at http://dnb.d-nb.de.
Any brand names and product names mentioned in this book are subject to trademark, brand or patent protection and are trademarks or registered trademarks of their respective holders. The use of brand names, product names, common names, trade names, product descriptions etc. even without a particular marking in this work is in no way to be construed to mean that such names may be regarded as unrestricted in respect of trademark and brand protection legislation and could thus be used by anyone.

Coverbild / Cover image: www.ingimage.com

Verlag / Publisher:
LAP LAMBERT Academic Publishing
ist ein Imprint der / is a trademark of
OmniScriptum GmbH & Co. KG
Heinrich-Böcking-Str. 6-8, 66121 Saarbrücken, Deutschland / Germany
Email: info@lap-publishing.com

Herstellung: siehe letzte Seite /
Printed at: see last page
ISBN: 978-3-659-80881-4

Contents

INTRODUCTION

For the span of millenniums the horse has been always at a man's side. It has been supporting him in his work and his feats of arms, has been providing him with food and clothes, has been serving truly in sport events, has been inspiring creators of folk epos. The horse has been always playing an important role in the cultural historic traditions of the nations populating Russia.

According to modern concepts the area of domestication of horses in 9500-2500 Millennium BC occupied a large part of the southern territories of modern Russia. This area stretched from the upper reaches of the Amur River to all the countries of Central and Southern Europe. In the past, the territory of Russia was the historic crossroads due to its geographic location, where for thousands of years the peoples of Asia and Europe moved, traded and fought that promoted an intensive process of horse breeds formation. That's why the study of molecular-genetic characteristics of the Russian population of horses is of interest to better understanding the processes of breed formations and the evolution of horse populations.

Now in Russia about 50 breeds of horses are being bred, including more than twenty indigenous breeds such as Orlov Trotter, Don, Kabardin, Altai, Bashkir, Mezen, Vyatka, Yakut and many others.

XV-XVI centuries saw the rise of Russian horse-breeding and there was a formation of stables and the studs. The development of trade relations with the West and the East contributed to the receipt of valuable horses of many breeds that were used in breeding. Imported horses had a considerable influence on the process of breed formation in the country. At the end of the XVIIIth century the Dutch, Danish, English and other European breeds along with the Arabian breed became the foundation for an excellent Orlov Trotter horse, which has retained its importance till now. Since that time Thoroughbred have begun to be bred in Russia and nowadays they constitute a special wealth for classical sports events and race and also serve as an indispensable improver in developing and improving so called riding half-bred lines.

The current breeds of Russian heavy draft horses trace from crosses of local mares with stallions of "cold-blooded" European breeds, such as Ardennas, Brabant, Percheron, Clydesdales, Suffolk and Shire. Although the need for the horses in agriculture has decreased, they still have a special role on small farms. In addition to their use for draught power, heavy horses have an important role in improving the meat quality of the local breeds.

The decline in the number of horses in modern conditions can result in the loss of unique rare populations and reduce genetic diversity of the species. According to the recommendations of the FAO (2007), the use of DNA markers is particularly important in the genetic monitoring programs for cultural and native populations, as well as rare and endangered species. Microsatellite markers are more likely than other methods to detect small differences between populations due to their high levels of allelic variation, being able to discriminate in both mean number of alleles and overall heterozygosity. Effective use of molecular genetic markers contributed greatly to the development of

3

software genetic-statistical analysis of populations (F-stat, PopGene32, Phylip, Tree, STRUCTURE). This made it possible to control the direction of genetic variability in populations, to study the phylogenetic relationships of breeds and to monitor the ongoing processes of microevolution.

The Laboratory of Genetics of the All-Union Research Institute for Horse Breeding was formed in 1980 and was accepted as an ISAG member in 1983. Currently it is the leading regional center for genetic control of horse parentage and study of the genetic characteristics of horses of different breeds and monitors the genetic diversity of populations.

In this book, a data set of 6810 horses was used to analyze genetic variability within and among 27 breeds of the panel microsatellite loci. I hope that the information provided will complement existing data on the characteristics STR polymorphism in domestic horse breeds of Europe and Asia.

Acknowledgements

This book was written thanks to many years of joint work of employees of the Laboratory of Genetics of the All-Russian Research Institute for Horse Breeding on the genetic testing of horses of different breeds. I thank my colleagues for their help in collecting samples and photos of horses of rare breeds and the information about the state of the breeds. I also thank Dr. Anna Ustyantseva for assistance in statistical analysis, interpreter Natalia Kiseleva and Julia Rozhnova for compiling this book.

Chapter 1. Horse breeds of Russia, their origin, characteristics and current status

Domestication and development of horse breeds is associated with Eurasian civilization. Russia has a significant part of the world resources of horse breeding. The State Registry of selection achievements includes 49 horse breeds and types, half of which are unique native populations. Russia has a long history of horse breeding, and various indigenous breeds were developed in different geographical regions. There were created such well-known breeds, as Orlov Trotter, Don, Budenny. Tersk, Altai, Bashkir, Kabardin, Vyatka and some other genetically distinctive populations in Russia.

1.1 Saddle breeds of horses

Breeds of riding horses were developed as a result of the intensive breeding. They are perfect in agility and other sport qualities, beautiful constitution and movements. These horses gained a high reputation in improvement horses of other breeds that are bred in many areas of the country and in the southern areas, in particular.

The *Akhal-Teke* is one of the oldest, most distinctive and unusual horse in the world. It is bred around the oases of the Turkmenistan Desert, north of Iran. Horses were bred and raced there 3,000 years ago. There is nothing in the world quite like this mystery horse. Its endurance and its resistance to heat are phenomenal. In 1935 Akhal-Tekes completed a ride from Ashkabad to Moscow, a distance of 4,152 km in 84 days. This extraordinary feat has never been equaled. Today the Akhal-Teke is a racehorse, a long-distance performer and a horse for sports, horse in the dressage and jumping disciplines.

Fig.1 The Akhal-Teke horse can be considered one of the oldest horse breeds.

Currently the horses of this breed are popular in many countries of Europe, Asia and America. The selection of Akhal-Teke horses is supervised by International Association of the Akhal-Teke Horse Breeding engaged at the All-Russian Research Institute for Horse Breeding. Only in the Russian Federation Akhal-Teke horses are bred at 3 studs, 12 breeding farms and by more than 100 private owners. More than 200 stallions and 650 mares are used for breeding purposes.

Arabian horse has emerged in Europe in the VIIIth century and has long been introduced to Russia. An extensive use of these horses in the XVIII-XIXth centuries resulted in the creation of our own highly valuable breeds such as the Orlov Trotter, the now extinct breeds Orlov Riding horse, Rostopchinskaya and Streletskaya. Besides, great attention is paid to breeding Arabian horses in pure.

Arabian horses are used to improve native breeds of Caucasus Region and Central Asia and also Don, Tersk and Trakehner breeds. The best Arabian horses were bred at Tersk Stud located in the Northern Caucasus. Progenitors of these horses were introduced to Russia from the best studs of England, France and Poland prior the Second World War and later on from Egypt.

Fig. 2 Russian bred Arabian horses have original type and high performance.

The *Thoroughbred horses* were introduced to Russia in the second half of the XVIIIth century. Volume I of the Russian Stud Book was issued in 1836; there were included 287 stallions and 366 mares. During the Soviet period the breed was perfected in comparative isolation. Since the mid 90-ies of the last century the country began to import stallions and mares from Europe and the USA. Only in the last ten years about 2.130 Thoroughbred horses were imported bfor racing and breeding.

Currently more than 1250 Thoroughbred mares are registered in the country. In Russia 15 studs and breeding farms are involved in breeding Thoroughbred horses.

Fig. 3 Thoroughbred horses race on 12 racecourses in different regions of Russia.

The **_Don_** breed is traditionally associated with the Don Cossacks, and evolved in the XVIIIth and the XIXth centuries. Its foundations were the steppe horses of the nomadic tribes. Early influences were the Mongolian Nagai and breeds like the Karabakh, the Persian Arab and the Turkmen.

Fig. 4 Don horses in the Rostov steppe

Don horses lived in herds on the steppe pastures and fended for themselves, scraping away the snow in winter to get the frozen grass beneath. Horses of Don breed are incredibly tough and adapts easily to every sort of climatic hardship.

Dons were then improved using Thoroughbreds and Arabians. The breed emerged as a solid army remount that could be put in harness, requiring minimum attention. Don horses are good-natured, calm and easily managed and quite able to work in harness and in light, agricultural draught. The Don is raced, mainly in long-distance events, and the present-day horse is larger and of better conformation than formerly. Don Horses were extensively used to develop the Budenny, Kustanair, New Kirghiz and Kushum breeds.

Fig. 5 The Budenny horses are used in sports and hikings

The **Budenny horse breed** was developed in the result of cross Don and native Chernomor mares with Thoroughbred stallions during the 1920-1930s. New breed was established in 1949. To improve athletic qualities mares were recrossed with Thoroughbred stallions. The Budenny horses are tested on racetracks and over long distances. On average the Budenny stallions stands 164 cm and has good jumping ability. The Budenny and Don horses were used to improve local steppe breeds of horses such as Bashkir, Zabaykalskaya, Kalmyk, Khakass and Tuva.

Fig. 6 Kabardin horses are well adapted to year-round keeping on mountain pastures of the North Caucasus.

The **Kabardin horse**, the breed of the Northern Caucasus, is derived from the horse of the steppe people crossed with Karabakh, Persian and Turkmen strains. This mountain horse, well-known since the XVIth century, is capable of working **in difficult terrain and is undeterred by snow and fast rivers.**

The Kabardin is at home in the mountains and due to developed characteristics that are suited to the terrain and the rigorous of the climate.

It is a sure-footed and agile horse. Predominant colours found in the breed are bay, dark bay and black without other distinguishing marking.

8

1.2 Trotter breeds

In Russia trotting is more popular than flat-racing. Currently on the Russian race tracks the representatives of the four following trotting breeds compete: Orlov, Russian, American and French trotters. Trotters have valuable qualities necessary for work in light harness, namely agility and high traction power. They are also appraised for their ability to genetically improve working horses raised on farms. Trotting at hippodromes enjoy great popularity.

Fig. 7 The Count Alexey Orlov and the stallion Bars I. Painting by N. E. Sverchkov (1871)

Orlov Trotter was created by the Count Alexey Orlov at Khrenovoe Stud at the end of the XVIIIth century. Using an Arabian, Dutch, Danish, English and many other European breeds he set out to create a perfect trotter. Selection of the best stallions and mares to be bred at the stud, the breeding in lines, efficient feeding – all this contributed to the creation and further improvement of the Orlov Trotter breed. Trotters tests held at different distances and selection on the fastest trotters used for breeding purposes made it possible to raise their worth.

By the XIXth century the Orlov Trotter breed became one of the most famous breeds in Russia that was kept and almost exclusively managed for a racetrack. The first Stud Book was published in 1847 and since that time the books have been regularly issued.

Until the eighties of the XIXth century, Orlov trotters were not only at home but also in Europe. Purely bred horses of prize type were exported abroad to participate in the run on race tracks in Germany, France, Italy and other countries. The best horses were used in a breeding of the French Trotter.

In the XXth century Orlov Trotter breed lost its leadership in playfulness to American Trotter. But in 1934 the absolute European record of 2 min 02.2 sec at

1600 m was set by the Orlov stallion Ulov. Modern absolute record of the breed of 1 min 57.2 sec was set by the stallion Cowboy in 1991.

Fig. 8 Lyrika Liubvi (Love You - Lotaringiya) is the fastest 2-year-old Russian Trotters

The *Russian Trotter* breed is the product of a long-term program of selective crossing of Orlov and imported American trotters that plays a competitive part in the sport. In the late nineteenth century American trotters were imported to many European countries, including Russia. That was the beginning of the work resulted in the development of a new breed – the Russian Trotter, established in 1949.

The past decades have seen an intensive process of crossing with the American Trotter breed that has a significant impact on the genetic structure of the breed. .The Russian Trotter holds many national records, like the 1600 m in 1 min 56.9 sec; 3,2 km in 4 min 06.1 sec.

Purebred breeding of *Standardbred horses* was started in Zlynsky stud in the 60-ies of the last century. Currently many horse owners are breeding American trotters so they are importing stallions and mares from Europe and USA for this purpose.

Breeding *French trotters* began in Russia relatively recently; this breed was included in the State Register of breeding achievements in 2009. French Trotter Versal won the Russian Derby in 2010 with the result 1 min 59.9 sec (1600 m); in 2013 the winner of this prize was Champion (2 min 01.0 sec). The leading farm breeding horses of this breed is Lokotskoy stud at Bryansk Region.

1.3 1.3 Heavy Draft horses

Currently there are bred mainly three native breeds of *Heavy Draft horses: Vladimir Draft, Russian Heavy Draft, Soviet Heavy Draft and Persheron*. Heavy draft horses perform different types of agricultural work on farms and are used to improve the quality of local breeds.

The Brabant, Ardennais, Percheron, Clydesdales, Shire and Suffolk breeds have been introduced from Western Europe since the second half of the XIXth century in the view of the industrial development and agricultural intensification. These heavy draft horses poorly adapted to continental climate conditions and were too slow at work over long distances, while their offspring resulted from the crosses with indigenous horses, did fit in with the requirements. In a subsequent period of time breeders would develop three new breeds of heavy draft horses by breeding their cross-breds.

Russian Heavy Draft horse breed is originated from old mountain-type

Ardennais horse that was introduced to Russia from Belgium in the second half of the XIXth century and was used for cross with native harness horses. An intensive breeding resulted in the development of a new more valuable breed was far more superior than the original Ardennais horse and at present it is sharply distinct from the horses that are bred in Western Europe nowadays. The breed belongs to the group of small draft horses. Measurements of a stud

Fig. 9 Russian Heavy Draft stallion Gigant.

stallion: height at withers - 152 cm, body length - 162 cm, girth - 206 cm, cannon`s circumference - 22 cm; measurements of a mare: 148-158-191-21. Coat colors are predominately chestnut, brown and red-roan.

Russian Heavy Draft horses are very popular among farmers because of their good adaptive qualities. They are successfully bred in different climatic zones, including Ural area and certain areas of Siberia.

Soviet Heavy Draft breed is based on the cross of Belgium Brabant and native harness mares. It was established in 1952. This horse is noted for its weighty body, bony legs, well-developed muscles. Measurements of a stallion: height in wither - 161 cm, body length - 169 cm, girth - 210 cm, cannon`s circumference - 25 cm; measurements of a mare: 159-166-200-24. Weight of stallion's averages 780 kg and of mares 650 kg. Mares of this breed are very peculiar for a high milk yield and in this respect no other breed can match. For example, the mare Ryabina produced

6.100 l of milk for a lactation period. Today the number of purebred mares of this breed makes 235. The most valuable Soviet Heavy Draft horses are at Mordovsky and Pochinkovski studs.

Vladimir Draft horse breed was developed on the basis of horse breeding in collective and state farms of Vladimir and Ivanovo Regions as a result of the cross of native mares with Clydesdale stallions. The breed was established in 1946.

At the exhibition Vladimir Draft horses attract many visitors by their stout build, bright coloring and concerted actions. Measurements of a stallion: 165-173-207- 24,5 cm, of a mare: 161–167–196–23,5 cm. The Vladimir breed is characteristic of a bay colour with white markings both on the forehead and legs.

Fig. 10 Vladimir Draft stallion Vernyi.

1.4. Native horse breeds

The populations of the native horses in European Russia have a wide range in the Northern, Central and Southern regions and are represented by such different breeds as *Bashkir, Vyatka, Mezen, Pechora* and others.

In the historical era in the Central and Southern regions of Russia one after another moving avalanche of nomadic peoples, horses which are inevitably included in the process of assimilation and participated in the formation of local breeds. Another important source of enriching the gene pool of populations was horses of Eastern origin, which came into the country through branches of the Great Silk Road. The *Bashkir horse* evolved centuries ago

Fig.11 Bashkir stallion Irandyk has a typical breed dun color with zebra markings on the legs.

12

in Bashkiria, around the southern foothills of the Ural Mountains. There it is bred as a pack, draft and riding animal that also provides meat, milk and clothing for people. The breed is well able to fend for itself in the most severe climatic conditions and is, in consequence, among the hardiest in the world. Bashkir horse has been crossed with the Dons and Budennys, and the latter, with both Trotters and Russian Draft stallions. Currently it is one of the most numerous native breeds of horses that are bred in several regions of Russia.

The *Vyatka horse* is an ancient native horse breed known since the XIVth century. The Vyatka horse is typical northern wood breed distinguished by good disease resistance, adaptive and exterior qualities. By the end of the XXth century the number of these animals was sharply reduced, and in 1996 Vyatka was given the status of disappearing breed (WWL-DAD:2).

Now areas of Vyatka horses have two nuclei - Udmurt and Kirov populations. By the first half of the XIXth century Vyatka horse was considered as the best troika horse in Russia.

A characteristic feature of the breed is its chestnut-roan or bay-roan colour with black stripe along the spine and wing- shaped patterns over the shoulders,

Fig. 12 Vyatka stallion Gertsog. Horses of this breed are used in agriculture, equestrian sports and driving.

as well as zebra stripes on the forelegs. However, the colour can also be brown, bay, chestnut or rarely black. The average measurements (in cm) of modern Vyatka mares are: height at withers 140, oblique body length 150, chest girth 172, and cannon bone girth 18,9. With the development of industry and transportation and with the intensification of agriculture, the numbers of Vyatka horses were sharply reduced; most of the purebred mares were mated with Heavy Drafts and Trotters. In recent years much attention is paid to the pure breeding of Vyatka horses and a lot of work is done for the conservation of this unique small breed.

Mesen horse is a typical representative of the northern forest breeds and lives in Arkhangelsk region on the shores of the White Sea. The breed has been known since the 16th century and was created the methods of natiural selection without significantly affecting the cultural species. Mesen horse has a harmonious addition and has the unique ability to move on the trot with a deep snow.

Fig. 13 Mezen horses systematically tested on the ability to carry loads in deep snow.

Fig. 14 Altai horse of the Ulagan district of the Altai Republic.

The local breeds of horses in Siberia were formed under the influence of Mongolian horses and had peculiar qualities permitting them to adapt well to harsh environmental conditions. The origin of these horses is lost in the remove antiquity. Judging by multiple equestrian bones found in burial mounds, the early nomads who inhabited this region in the first millennium B.C., had already had horses near to the modern local type. Herds of aborigine horses well apt to local conditions are grazed in the Altai Mountains, valleys of Tuva, Khakasia and Buryatia, Yakut taiga. The height at withes of the mares of the breeds averages from 135-138 cm, body length is 140-146 cm, girth is 164-172 cm, cannon's circumference – 17-17,5 cm. Indigenous people in many regions of Siberia willingly use horse meat in food, so the selection of local horses was conducted on the adaptive and meat quality.

In the Altai Mountains the locals have long been bred universal Altai horses. Altai horse carries the blood of the ancient Oriental breeds, which are moved together with the merchants on the Great Silk Road along the river Katun. Small horses of this breed are characterized by a wide variety of colors, among which the most popular is an appaloosa horse. Tuva horse lives in the foothills of the Altai and Sayan mountains and is well adapted to living in taiga, steppes and mountains. Thanks to this geographical isolation Tuva horse was formed without noticeable effect of other breeds. On separate farms it was used crossing of Tuva mare with Don and Budenny stallions which helped to improve meat quality of horses and to increase their agility.

Fig. 15 Tuva horses are able to maintain the fatness even on scant pastures.

In the far North in Yakutia people has been long breeding horses, exclusively adapted to extremely harsh climate conditions, when in winter the temperature reaches -70°C. Only Yakut horses are able to survive in the wild without shelter and to produce their own food from under the snow depth of 40-50 cm. A major portion of Yakut horses is distributed in the valleys of the middle part of the river Lena and also along the rivers Vilyui and Aldan; several herds are encountered further to the North, in the basin of the rivers Yana, Indigirka and Kolyma.

Yakut horses are not large, bone animals: mare's average height at withers varies with regions from 131 to 140 cm, body length ranges between 136 and 145 cm, girth – from 165 to 176 cm, cannon's circumference is 17,3-18,4 cm, weight is 370-430 kg. These typical forest horses or pony have a thick coat; the fat deposits under the skin contribute

Fig. 16 Yakut mare with foal is in the pasture.

15

to the preservation of heat in winter and   immunity to insect bites in summer. Yakut horses are raised for meat purpose and are also used under saddle for various works.

Yakut horse is the most numerous among local breeds and has several types. In the breed there are about 170 000 horses. Recently Prelenskaya and Megejecskaya horse breeds were formed in the result of mating of Yakut horses with Draft and Trotting horses.

The All-Russian Research Institute for Horse Breeding together with a number of regional agricultural institutions and a few Associations for different horse breeds make all efforts to study and conserve the unique genetic resources. There are issued stud books and breeding programs for all cultural horse breeds and for a number of local breeds of horses in accordance with the State Program "The development of horse breeding in the Russian Federation for 2013-2015 and for the planning period up to 2020"in the Institute. To encourage the development of horse breeding it is provided significant financial assistance to licensed studs and farms in the country.

Chapter 2. Polymorphism of microsatellite loci in the horse and its use or phylogenetic analysis of breeds

Microsatellites are a short tandem repeats (STR) with core sequences between 2 and 6 base pairs in length used in parentage testing, phylogenetic studies, population genetics and linkage mapping. Standard sets of microsatellites have been selected for population genetic studies and parentage assignment for livestock by the International Society for Animal Genetics (ISAG).

According to J Wright (1993), the loci of microsatellite DNA in its properties are universal genetic markers and can be used to solve many research problems. The loci of mini - and microsatellites in very large numbers scattered throughout the genome, allowing you to control the desired regions of chromosomes. These loci mainly localized in non-coding regions of the genome and, therefore, must be selectively neutral. This is a General rule, apparently, has exceptions in cases of close coupling with significant adaptive loci. In addition, although the function of mini - and microsatellites are unknown, a number of facts suggest that they can serve as coding or regulatory elements (Kashi , Soller, 1999).

For these loci are characterized by rapid evolution. The rate of spontaneous mutation of mini - and microsatellite loci is about 10^{-2} - 10^{-4} per locus per generation (Weber, Wong, 1993), which is much more than the allozyme genes – approximately 10^{-5}-10^{-6} (Nell et al., 1986; Nei, 1987). Therefore, if the divergence at allozyme genes (in the case of selective neutrality) is caused only by random drift, mini - and microsatellite loci and drift, and mutations (R. D. Ward, P. Grewe, 1994). Microsatellites are typical Mendel's signs with a codominant type of inheritance.

Using stepwise mutation model in the study of human populations showed that the mean estimated values of coefficients for the construction of a phylogenetic tree for a large number of microsatellite loci have an almost linear relationship with the duration of evolution. Such regularity due to the relatively high level of mutations in the regions of satellite DNA, which occur with a frequency of about 0,001 for one locus per generation (Nei M., N. Takezaki, 1996).

Equine STR loci are extensively used for parentage verification by the horse breeding industry. The first description of horse STR loci in the 1990s has eventually led to international recommendation for these loci in 1998 by the International Society for Animal Genetics. The ISAG currently recognized set of 12 STR loci (AHT4,

AHT5, ASB2, ASB17, ASB23, HMS2, HMS3, HMS6, HTG4, HTG7, HTG10 and VHL20) for use in equine kinship analysis and has validated these in numerous inter-laboratory comparison test (Van de Goor et al., 2011). For paternity testing and parentage verification laboratories usually used the current 17-plex microsatellite kit for genotyping horses (Table 1).

Table 1 Locus name, number of alleles, chromosomal location, amplicon length, primer sequence and original reference STR loci

Locus	Number alleles*	Chromosome location	Amplicon length (bp)	Primer sequences (5'-3')	Original reference
AHT4	11	24q14	140-166	F: aaccgcctgagcaaggaagt R: cccagagagtttaccct	Binns et al, 1995
AHT5	11	8	126-147	F: acggacacatccctgcctgc R: gcaggctaagggggctcagc	Binns et al, 1995
ASB2	16	15q21.3-q23	237-268	F: ccttccgtagtttaagcttctg R: cacaactgagttctctgatagg	Breen et al, 1997
ASB17	15-19	2p14-p15	104-116	F: gagggcggtacctttgtacc R: accagtcaggatctccaccg	Breen et al, 1995
ASB23	9-14	3q22.1-q22.3	176-212	F: gaggtttgtaattggaatg R: gagaagtcatttttaacacct	Breen et al, 1995
CA425	11	28	224-247	F: agctgcctcgttaattca R: ctcatgtccgcttgtctc	Der Valle A. et al, 1997
HMS1	8	15	166-178	F: catcactcttcatgtctgcttgg R:ttgacataaatgcttatcctatggc	Guerin et al, 1994
HMS2	12	10	218-238/	F: acggtggcaactgccaaggaag R: cttgcagtcgaatgtgtattaaatg	Guerin et al, 1994
HMS3	13	9	146-170	F: ccaactctttgtcacataacaaga R: ccatcctcactttttcactttgtt	Guerin et al, 1994
HMS6	8	4	154-170	F: gaagctgccagtattcaaccattg R: ctccatcttgtgaagtgtaactca	Guerin et al, 1994
HMS7	11	1q25	167-186	F: caggaaactcatgttgataccatc R: gttgttgaaacataccttgactgt	Guerin et al, 1994
HTG4	12	9	116-137	F: ctatctcagtcttcattgcaggac R: ctccctccctccctctgttctc	Ellegren et.al, 1992
HTG6	13	15q26-q27	73-103	F: cctgcttggaggctgtgataagat R: gttcactgaatgtcaaattctgct	Ellegren et.al, 1992
HTG7	8	4	114-126	F: cctgaagcagaacatccctccttg R: taaagtgtctgggcagagctgct	Marklund et al, 1994
HTG10	14	21	83-105	F: caattcccgccccacccccggca R:tttttattctgatctgtcacattt	Marklund et al, 1994
LEX3	14	Xq	137-160	F: acatctaaccagtgctgagact R: aaggaaaaaaaggaggaagac	Coogle L.et al, 1996
VHL20	12	30	83-102	F: caagtcctcttacttgaagactag R: aactcagggagaatcttcctcag	Van Haeringen et al, 1994

* Data by Van de Goor et al. (2010).

The undeniable advantage of molecular genetic markers is the possibility of their use for studying phylogenetic relationships and evolution of related species (Breen et al., 1994; Folch et al., 1996; Kruger et al., 2002). M Breen et al. (1994) demonstrated that equine microsatellites can be used for genotyping of the genus Equus, in particular, the Przewalski horse. Comparative analysis of polymorphism at the horse and two species of Zebra (Y. Moodley et al., 2006) showed that the representatives of the wild equine have a fairly high level of genetic diversity.

L.H.P. van de Goor et al (2010) have presented data for comprehensive panel of 35 equine populations on a selection of loci used routinely by several laboratories (Table 2).

Table 2 Conversion of ISAG nomenclature to repeat-number nomenclature (Van de Goor et al., 2010)

Repeat number	ISAG nomenclature STR loci																
	AHT4	AHT5	ASB2	ASB17	ASB23	CA425	HMS1	HMS2	HMS3	HMS6	HMS7	HTG4	HTG6	HTG7	HTG10	LEX3	VHL20
9		B															
10		C															
11			D														
12				D	F								G				
13		F	F		G					K	G					F	I
14	H	G	G		H	I				L			I			G	J
15	I		H	G	I	J	H			M			J	K		H	K
16	J	I	I	H	J	K	I			N	J		K		H	I	L
17	K	J	J	I	K	L	J			O	K			M	I	J	M
18	L	K	K	J	L	M	K			P	L		M	N	J	K	N
19	M	L	L	K	M	N	L			Q	M		N	O	K	L	O
20	N	M	M	L	N		M	H			N		O	P	L	M	P
21	O	N	N	M	O			I			O		P	Q	M	N	Q
22		O	O	N	P						P		Q		N	O	R
23		P	P	O	Q						Q				O	P	
24		Q	Q	P					L						P	Q	
25	H		R	Q					M						Q		
26	I		S	R					N						R		
27	J		T	S					O						S		
28	K		U	T					P						T		
29	L		V	U					Q								
30	M		W	V					R			K					

19

31	N						S		L				
32	O		Y						M				
33	P								N				
34	Q								O				
35	R								P				
36									Q				

Population studies were performed (L.H.P. van de Goor et al 2011) for 17 panel STR loci including 8641 horses representing 35 populations. Several important breeds were not included (e.g. Akhal-Teke, Clydesdale, Exmoor, Morgan, Mustang and Noriker). Rsearchers demonstrate the phylogenetic signal of the 17 STR and found three clusters of related breeds: (i) the cold-blooded draught breeds Haflinger, Dutch draft and Friesian; (ii) the pony breeds Shetland and Miniature horse with the Falabella, Appaloosa and Iceland; and (iii) the Warmblood riding breeds, together with the Standardbred, Thoroughbred and Arabian.

Using 27 microsatellite loci Y.H. Ling et al (2010) determined the genetic diversity and evolutionary relationships among 26 Chinese horse breeds come from 12 different provinces. Although there were abundant genetic variations found, the genetic differentiation was low between the Chinese horses, which displayed only 2,4% of the total genetic variance among the different breeds. However, genetic differentiation (pairwise FST) among Chinese horses was still apparent and varied from 0,001 for Guizou-Luoping pair to 0,064 for the Jingjang-Elenchuns pair. Genetic relationships among Chinese horse breeds were also consistent with their geographical distribution. The Thoroughbred and Mongolia breeds could be discerned as two distinct breeds, but Mongolia horse in particular suffered genetic admixture with Chinese horses.

C. Luis et al (2007) studied diversity and genetic relationships of three native Portuguese horse breeds and other horse populations from Asia, Europe and America based on protein and microsatellite loci variation. The combined 17 blood systems and 12 STR data produced a tree that fit historical records well and with greater confidence levels than those for either data set alone. Trees obtained for each of markers analyzed (microsatellite, protein and blood groups) showed some cluster differences. However, there were some consistent 4 groups in all trees: (i) American Saddlebred, Rocky Mountain, Standardbred and Morgan Horse; (ii) Friesian, Dales Pony and Fell Pony; (iii) Lusitano and Andalusian; and (iv) Irish Draught, Quarter Horse, Hanoverian, Thoroughbred and Holstein.

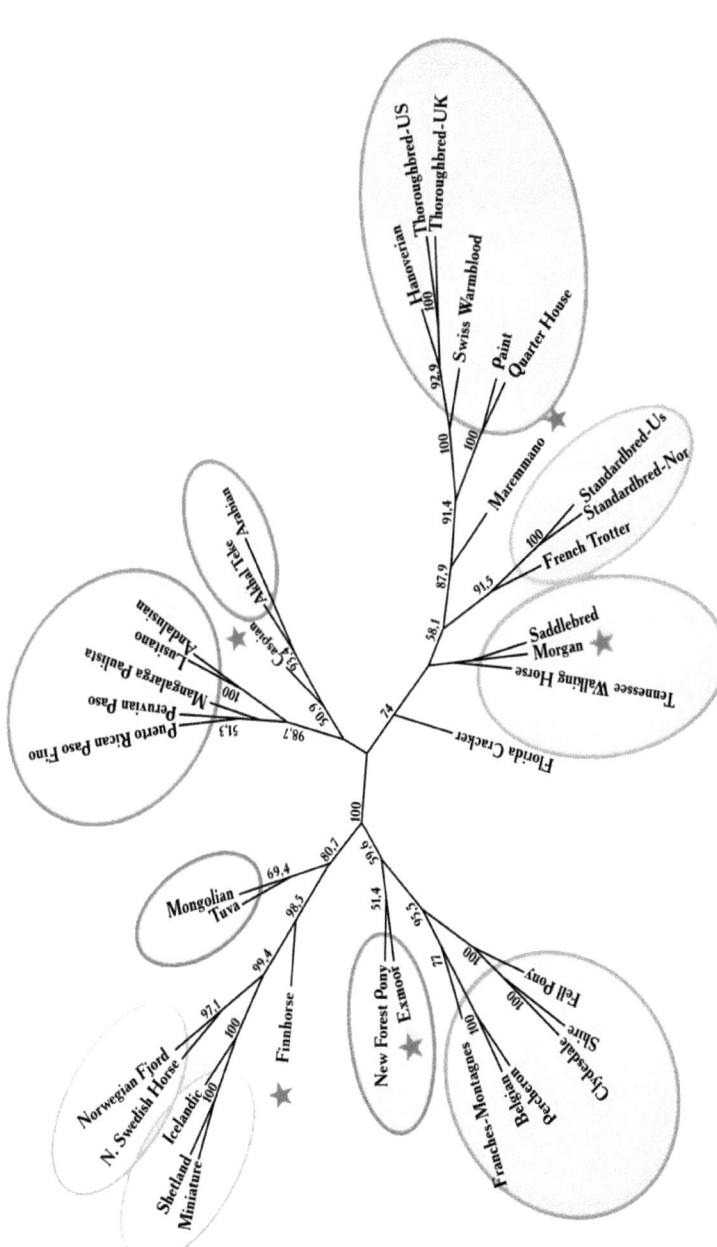

Fig. 17 Dendrogram of genetic relationships between horse breeds based on genome-wide analysis (Mickelson et al., 2012)

21

One of the clusters included the Caspian Pony, Arabian, Akhal-Teke and Lipizzaner breeds. The first tree breeds are part of the oriental-type horses, so this clustering would be expected based upon their origin in the Middle East and Asian steppes (Cothran, Luis, 2005).

Mickelson et all (2012) are using genome-wide SNP genotype data collected from greater than 20 horses from 33 breed at over 54,000 loci to attempt to identity some of these genomic regions under selection. Analyses show substantial variability in genetic diversity amongst breeds and demonstrated relationships among the breeds that largely reflect geographic origins and known breed histories.

Numerous publications suggest that the genetic structure of many horse breeds from different countries at STR loci have been well studied, but there is little information about the genetic features of the original Russian horse breeds. Therefore, the next Chapter will present the results of the study of polymorphism of microsatellite loci in horses cultural and local breeds, bred in Russia.

Chapter 3 Microsatellite diversity in Russian horse breeds

Genetic diversity between 27 horse breeds bred in Russia was analyzed using 17 panel microsatellite markers. High levels of polymorphism were observed in populations with the mean number of alleles (MNA) and average of observed heterozygosity (Ho) at 6.47 and 0.695, respectively. The number of alleles at loci varied between 2 (HTG6) and 15 (ASB17).

In addition to the standardized DNA allele's nomenclature of 17 equine-specific STR loci were found another alleles (AHT5P, ASB2H, ASAB17U, ASB23M, ASB23N, CA425E, HMS1O, HMS1R, HMS2D, HMS2G, HMS2N, LEX3R and LEX3S) in indigenous horse breeds. The private alleles were registered only in some indigenous breeds including Altai, Bashkir, Kabardin, Mezen, Pechora, Vyatka and Yakut.

3.1 Genetic variability of microsatellite loci in the horse riding breeds

The study of genetic diversity of alleles of STR loci in horses of Akhal-Teke, Arabian and Thoroughbred breeds is of particular interest, as these three breeds had a huge impact on the world horse breeding. Comparative analysis of polymorphism of 17 microsatellite loci in horses of these three breeds show that the oldest Akhal-Teke horse breed has the widest range of alleles, total 130, that is a record for horses of stud and local breeds of our country. Horses of these three purebred breeds have specific allele spectrum of satellite DNA for 16 loci and similarity has been found only in locus HMS7 (Table 3).

For genetic and population analysis there were used the results of DNA typing of 2024 Akhal-Teke horses, issued in the form of DNA-certificates by the genetic laboratories of different countries, including Italy, Germany, France, Czech Republic, Russia, USA and Turkmenistan. The numbers of alleles at loci were high and ranged from 4 to 12, the loci ASB17 and HMS2 showed the highest polymorphism.

22 alleles of 17 STR loci not found in Arabian and Thoroughbred horses were typed in Akhal-Teke horses, as well as several private alleles, ANT5Q, HMS2N and HMS2T, which have not been detected in horses of other breeds. In the locus LEX3 there were determined 9 of the 14 known alleles that indicates a fairly wide parent the maternal foundation of this breed.

It is necessary to note a certain parallel between the individual rare alleles of Akhal-Teke horses and local horses of Siberia, which indicates the relationship of these populations of horses, and, perhaps, common ancestors in the distant past. For example, alleles ASB17F and ASB17T, and CA425I were identified only in the genotypes of the Akhal-Teke, Tuva and Khakass horses.

Comparison between 2024 Akhal-Teke horses from different regions demonstrate that each population has allele pool typical for this breed. Genetic characteristics of the populations varied within a narrow range (Table 4). A wide spectrum of alleles, including rare alleles ASB17F и ASB17T, HMS1K, HMS3H, HMS6N and HTG7P was found in Turkmenistan horses (n=410). This population of Akhal-Teke horses is characterized by high level of genetic variation (Na=121) and consolidation (F=0,030). Coefficients of genetic similarity demonstrate more close relationship between Russian and European populations of Akhal-Teke horses (0,987) and a more significant differentiation between Turkmenistan and American horses (0,954). Maximum index of differentiation is observed in the Akhal-Tekes of America (Table 4).

Table 3 The range of alleles of 17 microsatellite loci in Akhal Teke, Arabian and Thoroughbred horses

Loci	No alleles in locus	Breed of horses					
		Akhal-Teke n=2024		Arabian n=420		Thoroughbred n=1945	
		Typical alleles p > 0,05	Rare alleles q < 0,05	Typical alleles p > 0,05	Rare alleles q < 0,05	Typical alleles p > 0,05	Rare alleles q < 0,05
AHT4	11	H, J, O	I, K, M, N	H, J, K, M, O	I, P	H, J, K, O	I
AHT5	11	J, K, N, O	L, M, Q*	J, K, M, N,	O	J, K, M, N,	O
ASB2	16	I, K, M, N, P, Q, R	B, O	B, J, K, N, O, Q	C, I, L, M, R	B, K, M, N, O, Q, R	I, P
ASB17	15-19	G, H, N, Q, R, S	F, K, M, O, P, T	N, O, R	G, K, M, Q	G, M, N, O, R	Q
ASB23	9-14	I, J, K, L, U	G, R, S	I, J, K, L, S	U	I, J, K, L, S	U
CA425	11	J, M, N, O	I, K, L	J, N, O	I, L, M	I, J, N, O	K, L, M
HMS1	8	I, J, M	K, L, N	I, J, L, M		I, J, M	L
HMS2	12	H, I, K, L, M, P, R	J, N*, O, T*	H, L, M, P	I, K, R	H, J, K, L, M	I, P

24

HMS3	13	I, M, N, O, P, Q	H, R, S	I, M, N, O, P	Q	I, M, O, P	N, R
HMS6	8	L, M, O, P	K, N	L, P	K, M, O	K, M, P	L, O
HMS7	11	J, K, L, M, N, O,	-	J, K, L, M, N	O	J, L, M, N, O	K
HTG4	12	K, M, P, Q	L, N, O	K, M, L, N,	P	K, M	L, N, O, P
HTG6	13	G, J, O	P	G, J, O	-	G, J, O	*M, R*
HTG7	8	K, N, O	M, P	K, N, O	-	K, N, O	M
HTG10	14	K, L, O, R	I, M, P, Q	I, K, L, O, R	M, N, S	I, K, L, M, O, R	S
LEX3	14	F, H, L, M, N, O, P	I, K	F, H, I, M, P	K, L, N, O	H, M, N, O, P	F, I, L
VHL20	12	J, M, N, O, Q, R	I, L, P	I, L, M, N, R	P	I, L, M, N	O, R

Table 4 Genetic characteristics of the different populations of Akhal-Teke horses on 17 STRs loci

Parameters	Russia	Turkmenistan	Europe	CIS*	America	Total
Number of horses	787	410	440	254	133	2024
Number of aleles	117	121	119	117	110	130
Number of alleles per locus	6,88	7,12	7,00	6,88	6,47	7,65
Ae	3,72	4,01	3,71	3,86	3,71	3,87
Ho (%)	68,65	71,45	68,80	70,51	71,49	69,58
He (%)	68,50	71,36	68,75	69,17	68,73	69,70
Fis	-0,002	-0,001	-0,001	-0,020	-0,039	0,001
Fst	0,019	0,030	0,014	0,011	0,014	-
I_D – differentiation index	0,094	0,173	0.128	0,150	0,262	-

*CIS - The Commonwealth of Independent States (Azerbaijan , Armenia, Kazakhstan, etc.).

Akhal-Teke horses from different regions of Europe and Asia have similar genetic characteristics (Ae, Ho, He, F_{IS}, F_{ST} and I_D) and high degree of consolidation. High polymorphism of microsatellite, blood group and biochemical markers in Akhal- Teke horses proves the ancient origin of the breed (Ryabova et al., 2012).

All of the regional populations of Akhal-Teke horses have a typical genetic structure and it is available rather large variability of resource for the successful development of this breed that is evidenced by Szontagh et al (2005).

In Arabian and Thoroughbred horses 104 and 102 alleles were found, respectively. From 17 STR loci only ASB2 and LEX3 are highly polymorphic and three (HMS1, HTG6 and HTG7) present low polymorphism. Genetic distances between these breeds demonstrate more close relationship between Arabian and Akhal-Teke (0,271) and less significant similarity between Arabian and Thoroughbreds (0,275).

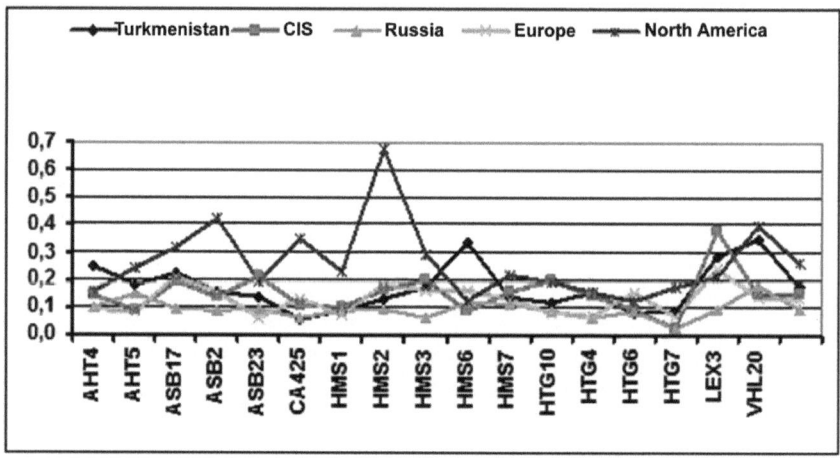

Fig. 18 The level of genetic differentiation (ID) of Akhal-Teke horses from different regions.

The Russian population of Arabian horses, traced to Tersk stud, is characterized by high consolidation (Ae 3,05; Ho 0,616). The genotypes of Arabian horses are determined by 3 alleles (ASB2C, ASB2J and HTGN) that are absent in Akhal-Teke and Thoroughbred horses. The genetic structure of the Arabian breed is characterized by high frequency of alleles AHT5 N (0,561), ASB2Q (0,548), ASB17R (0,456), HMS2M (0,264), HMS6P (0,741), HTG7O

26

(0,752), HTG10L (0, 426), CA425N (0,732) and VLH20L (0,425). Imported from France the Arabian sires Benedick and Nougatin enriched the allele pool of the breed with several rare alleles including AHT4I, AHT4P, AHT5O, HMS2K, HMS6O and HTG10N.

The study of the genetic characteristics of the genealogical structure of Arabian horses shows certain differences between lines and mare families on the number and frequency of alleles and degree of heterozygosity of microsatellite loci (Zaitceva et al., 2010) that are supported by a system of selection even at a fairly high level of inbreeding (F_x =3,72%). The highest level of genetic diversity is in the line of Latif (Ae=3,36; Ho=0,693), lower Ae values (2,94-2,50) are in the lines of Koheilan I, Korej, Mansour and Naseem.

The coefficients of genetic similarity between lines vary in the range of 0,820- 0,958. Minimal differences in microsatellite markers were found between the lines of Koheilan 1 and Korej (0,04), Koheilan 1 and Mansour (0,05) and Mansour and Korej (0,06). The most varied were the lines of Amurath and Latif (0,200) that is clearly demonstrated in the dendrogram (Fig.20).

Cluster analysis show the presence of intra-breed genetic differences between male and female lines even in a small population of Arabian horses, which indicates the promising use of STR's markers in breeding programs.

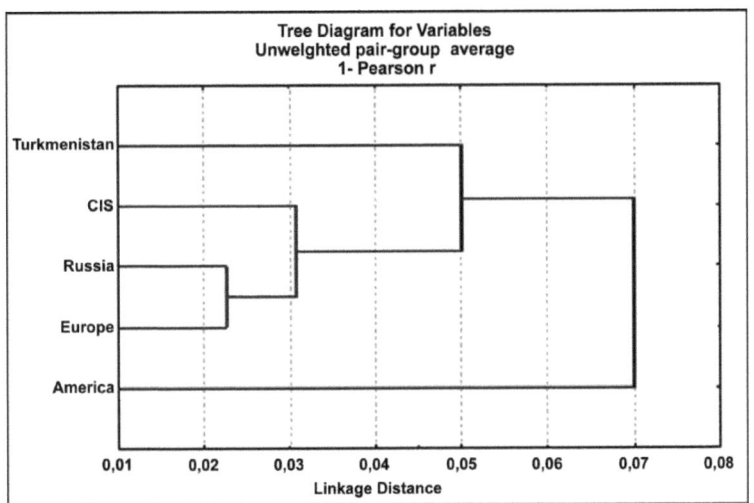

Fig. 19 The dendrogram of genetic distances of Akhal-Teke horses from different regions.

Analysis of the results of genotyping of 1945 Thoroughbred horses show that the Russian population is typical for the breed spectrum and the structure of alleles of the all 17 microsatellite loci. Unlike the horses of Akhal-Teke and Arabian breeds, in Thoroughbred horses only two additional minor allele were found (HTG6M and HTG6R). Both alleles with low frequency (0.003-0,009) are met in Thoroughbred horses of domestic as well as of foreign origin.

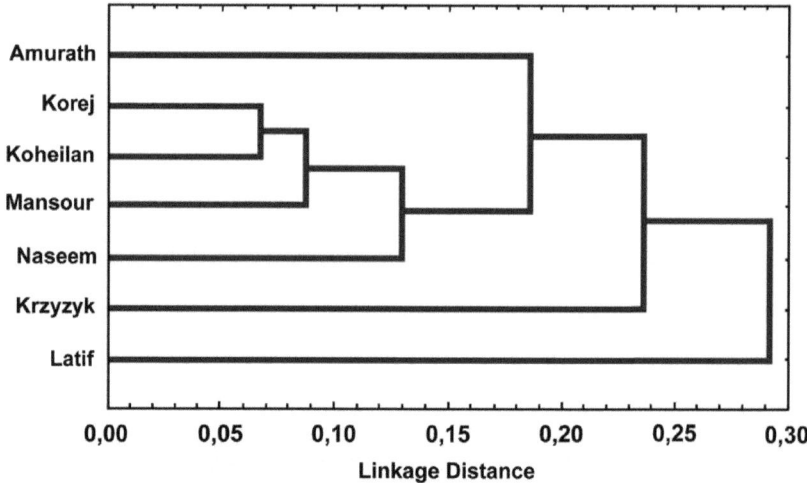

Fig. 20 The dendrogram of genetic distances between different lines of Arabian horses at STR loci

Genetic structure of Thoroughbred horses is characterized by a high frequency of alleles ASB17G (0,315), HMS2L (0,655), HMS3I (0,566), HMS6P (0,573), HTG4K (0,492), HTG6J (0,462), HTG10I (0,398), LEX3P (0,306) and VHL20N (0,312). The observed heterozygosity (Ho) resulted in 0,696 and ranged between 0,561 (HTG4) and 0,830 (ASB2).

To explore the impact of imported Thoroughbred horses on the allele pool of the Russian population it was carried out a comparative analysis of the genetic characteristics of stallions and mares of different origin. In the genotypes of 69 imported stallions 64 alleles were determined in 13 microsatellite loci, whereas in 113 domestic sires variety of options was slightly higher and reached 69. But the differences in the spectrum of alleles of these two groups of stallions were due to presence of very rare variants

(HMS7K, HTG4O, HTG6M, HTG6R, VHL20O), that are met with a frequency of 0,005-0,025. The group of imported stallions had a slightly higher level of polymorphism, but their heterozygosity was significantly lower than in domestic sires (Tab. 5).

Thoroughbred mares born in our country and imported were characterized by almost the same level of genetic diversity. Imported mares enriched the allelic spectrum of the national population Thoroughbred horses with new alleles HMS6O and HTG10S. The mares from other countries had a significantly higher frequency of alleles AHT5 K, ASB2 Q, HMS7J and HTG4K (P>0,999).

Table 5 Characteristics of Thoroughbred stallions and mares at STR loci

Groups of horses	n	Ae	Ho	He	Fis	I_D
Sires imported	69	3,35	0,656	0,670	0,021	0,174
Sires bred in Russia	113	3,33	0,735	0,750	0,020	0,141
Sires, total	182	3,42	0,679	0,693	0,020	0,081
Mares imported	87	3,30	0,652	0,661	0,014	0,209
Mares bred in Russia	473	3,28	0,650	0,666	0,024	0,055
Mares, total	560	3,32	0.656	0,670	0,021	0.031

In almost identical number of effective alleles of microsatellite loci bred in Russia Thoroughbred stallions are characterized by a higher degree of heterozygosity (Ho) compared to imported sires, while mares of Russian and foreign selection have the same degree of heterozygosity. In all compared groups of horses, there is a slightly positive value of Fi8, which indicates an excess of horses with homozygous genotypes.

Genetic differences STR loci between the Russian population of Thoroughbred horses and the Thoroughbreds of foreign selection are illustrated by the dendrogram (Fig. 21), which clearly shows two different subclusters. Imported mares are the most genetically heterogeneous group in comparison with studied population of Thoroughbreds, as evidenced by the largest index of differentiation (ID =0,209).

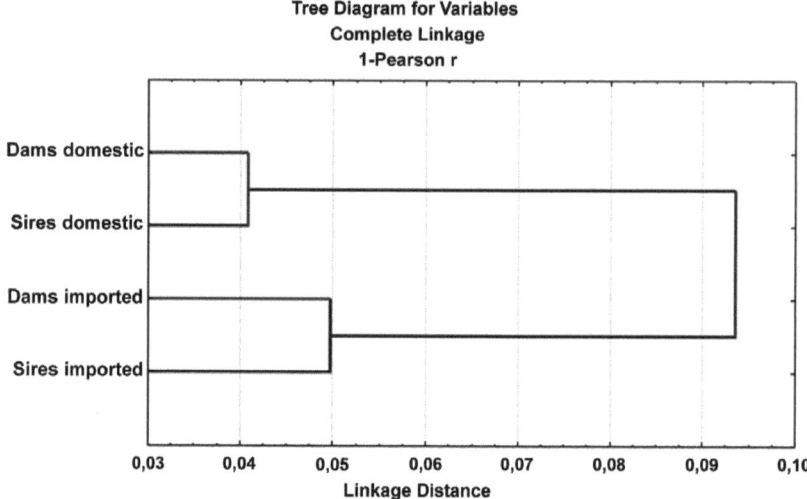

Fig. 21 The dendrogram of genetic distances of Thoroughbred stallions and mares of different origin.

Genotyping results of 182 stallions have been analyzed to assess the degree of differentiation of the linear structure of English Thoroughbred horses at molecular-genetic level. The greatest diversity of alleles and types of 13 microsatellite loci were identified in stallions of the most numerous lines of Native Dancer (62), Northern Dancer (61), and Nasrulla (61). The stallions of different lines differ on presence and frequencies of individual alleles and types of microsatellite loci, as well as on the level of polymorphism (Ae) and the degree of heterozygosity (Ho) as some loci, and their averages. The average degree of heterozygosity (Ho) in different lines of Thoroughbred stallions varied in the range of 0,665-0,716, but interline variability of heterozygosity of the loci was very high (Fig. 22).

A high degree of diversity of genotypes of sires of different lines was registered at the loci AHT4 (2,42–3,79), HMS7(2,04-4,38), HTG10 (2,67–5,14) and VHL20 (1,41-3,88). But even in loci with a low level of polymorphism linear differences were noticeable: HMS1 (1,44-2,88), HMS2 (1,36 - 3.60) and HTG6 (1,51-3,27). The analysis of the degree of genetic differentiation of the lines of Thoroughbred breed by microsatellite DNA loci showed that the highest index of differentiation and deviations from average values are in representatives of the line

30

of Massine (Fr). Stallions of Douglas line, formed in Russia, had high differentiation level (I_D =0,431).

The coefficients of genetic similarity between lines varied in a wide range 0,791- 0,960, with the highest similarities between stallions of Nasrulla and the Northern Dancer lines that trace to Phalaris.

Interestingly, the dendrogram of genetic distances between lines to a large extent coincides with modern genealogical structure of the Thoroughbred breed. The dendrogram clearly show the central cluster that combines all the four branches (Nasrulla, Northern Dancer, Nearco, Native Dancer) of the old line of Phalaris. Certain genetic relationship is seen between the lines of Douglas and Massine, ascending through Florizel II to the prominent sire St. Simon (1891). A bit isolated line of Man O`War was presented by direct male descendants of Matchem, one of the founders of the Thoroughbred breed.

Comparative assessment of the level of polymorphism of microsatellite loci in Thoroughbred mares of different families showed that there are also marked differences in the number of effective alleles (2,64-3,20) between them. Relatively high variability of the level of polymorphism was observed in the loci ASB2 (3,76-6.46), HMS7 (2,55-5,45) and HTG10 (2,61-4,70). The coefficients of genetic similarity between Thoroughbred female lines ranged in the interval 0,792-0,957. The index of differentiation of the female lines, characterizing the difference from the average for the breed, ranged from 0,196 to 0,469.

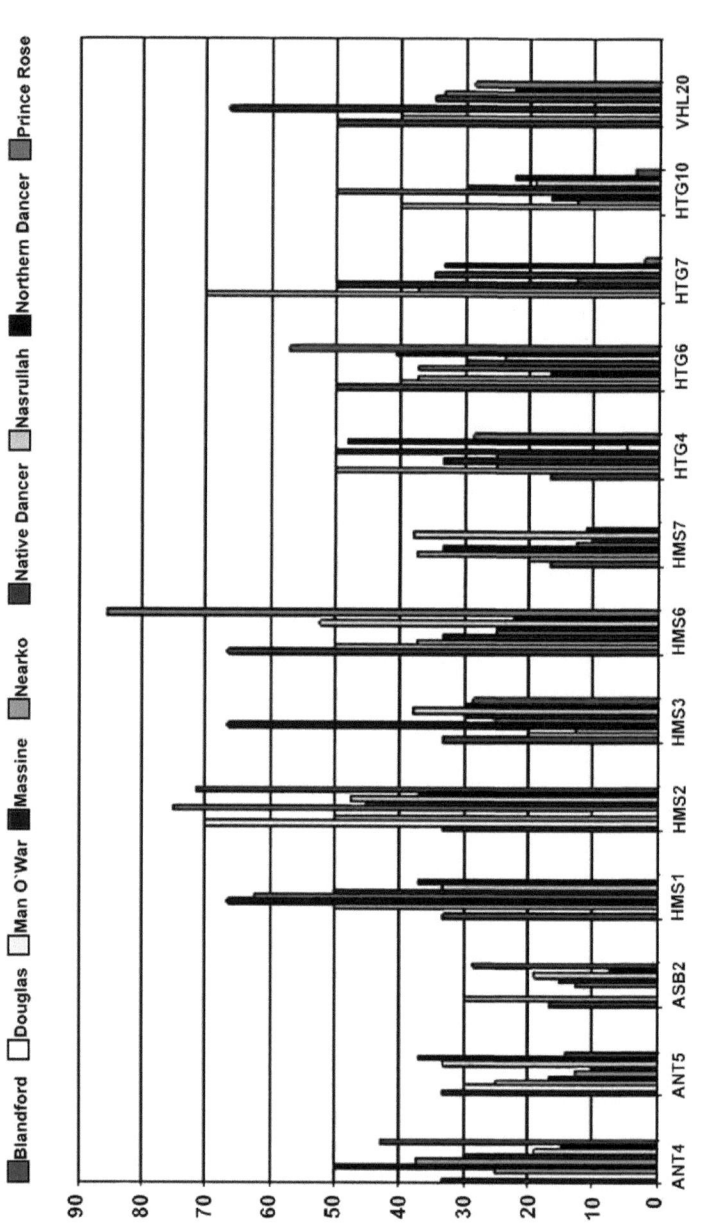

Fig. 22 Interline variability of observed heterozygosity (Ho) STR loci in Thoroughbred stallions.

32

The analysis of molecular-genetic features of the Thoroughbred male and female lines indicates the existence of sufficient genetic differentiation of the genealogical structure of the domestic population of Thoroughbred horses on the spectrum and frequencies of alleles of satellite DNA. Breeding lines contribute to the formation and consolidation of valuable genetic complexes that, in turn, lead to breeding heterosis and progressive development of the breed.

A total of 99 alleles were typed at 17 microsatellite loci in horses of Budenny breed. Allele spectrum of horses of this breed is shown in Table 6. Unlike horses of three previously described breeds in Budenny horses were identified additional rare alleles AHT4L, AHT5P, ASB17D, ASB23T and HTG6I The average number of allele was 5,82 which ranged from 3 to 9 per locus. The observed heterozygosity (Ho) ranged from 0,388 to 0,944 (the average value was 0,654).

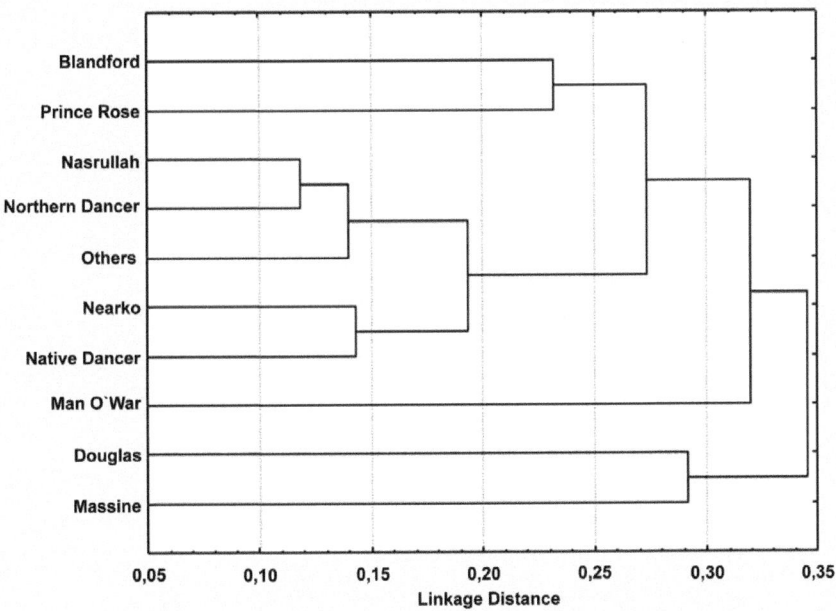

Fig. 23 The dendrogram of genetic distances between Thoroughbred stallions of the different lines.

In Don Horses fairly high level of genetic diversity of microsatellite loci was determined, given the small number of tested animals (n=17). The average number of allele was 5,53 which ranged from 3 to 10 per locus. The effective number of alleles (Ae) was 3,39. The highest allele frequency in each locus was 0,353 for AHT4K, ASB2M and HMS3, 0,588 for AHT5N, 0,529 for HMS2N, 0,412 for HMS6M, 0,382 for HTG4, 0,736 for HTG7, 0,383 for HTG10R, 0,324 for VHL20I and VHL20M 0,412 for ASB23J, ASB17N and LEXH.

Only in the Akhal-Teke and Don horses unique allele ASB17H was revealed that confirms the presence of the oriental blood in this steppe breed.

By genetic analysis, it was found that horses of Budenny breed have a higher coefficient of genetic similarity (0,896) with Thoroughbred than with the Dons (0,733). Horses of Don and Thoroughbred breeds have a low level of genetic similarity (0,661), which may be due to the originality and uniqueness of the allele pool of Don horse bearing the genes of many domestic breeds.

Table 6 The range of alleles of 17 STRs loci in Budenny, Don and Tersk horse breeds

Loci	No of alleles in locus	Breed of horses					
		Budenny n=33		Don* n=17		Tersk n=15	
		Typical alleles p > 0,05	Rare alleles q < 0,05	Typical alleles p > 0,05	Rare alleles q < 0,05	Typical alleles p > 0,05	Rare alleles q < 0,05
AHT4	11	H, J, K, O, P	L, M	H, J, O, P	K, Q	H, J, O, P	
AHT5	11	J, K, M, N	L, O, P	K, M, N		J, K, M, N	
ASB2	16	B, K, M, N, O, P, Q, R	I	K, M, N, O, P	I, Q	B, K, M, O, P Q	
ASB17	15-19	G, M, N, O, P, R	D, L, T	G, H, K, L, N, R, T	M, P, Q	G, K, M, N, R, S	Q
ASB23	9-14	I, J, K, L,	S, T	G, H, J, L, S, U	K	I, J, K, S	
CA425	11	J, N, O	M	-	-	J, M, N, O	
HMS1	8	I, J, M		-	-	I, J, M	
HMS2	12	I, J, K, L, M	H	J, K, M, N	O	H, L, M, R	I, P
HMS3	13	I, M, N, O, P		I, M, N, O, P, R, S	Q	M, O, P	
HMS6	8	L, M, P	K, O	L, M, P	O	K, M, P	L

HMS7	11	J, L, M, N, O	K	J, K, L, M, O		J, K, L, M, O	N
HTG4	12	K, M, N	O, P	K, L, M, P		K, M	N, O
HTG6	13	G, J, M, O	I	G, J, O		G, J, O	
HTG7	8	K, N, O		K, N, O		K, N, O	
HTG10	14	I, L, M, O, R	K, Q	I, M, O, R	K, L	I, K, L, M, O, R	
LEX3	14	H, L, M, N, O, P		F, H, L, O, P	N	H, L, M, O, P	
VHL20	12	I, L, M, N, R	O	I, L, M, N, P, R		I, L, N, P	M

Note * results DNA testing of the Don horses provided by E. G. Cothran.
A_e – effective number of alleles; H_e –expected heterozygosity;
H_o – observed heterozygosity F_{is} – population inbreeding level; MNA- mean number alleles per locus.
* Don horses were tested at 15 STR loci.

A total of 77 alleles at 17 loci were detected in Tersk horses (n=15), with locus ASB17 showing the highest number of alleles (Table 6). This endangered horse breed with the Anglo-Arab foundation was characterized by high frequency of the following alleles: VHL20L (0,400), HTG4M (0,834), HTG6 (0,867), ASB2Q (0,500), HTG10K (0,500), HMS2L (0,600), ASB17S (0,380), HMS1 (0,733) and CA425M (0,572). In the genotypes of Tersk horses rare alleles HMS2P and HMS2R were registered that were not met at Thoroughbred and Arabian horses and may had been inherited from Strelets horse breed.

The A_e and H_o values presented in Table 7 indicate a low level of genetic diversity in Tersk breed of horses. Perhaps this is due to the limited number of horses that participated in the creation of this breed, which inevitably led to the use of inbreeding.

A comprehensive study of the genome of Kabardin horse breed that is performed with the support of the German Foundation "VolkswagenStiftung", revealed high level of genetic diversity allelic variants of microsatellite loci in the studied samples (n=303). A total of 165 alleles were detected at 17 microsatellite markers (Duduev et al., 2014). The average number of alleles per locus was 9,7 and ranged from 6 (HTG7) to 15 (ASB17). The genetic analysis identified rare alleles of microsatellite loci that were not described in horses of European breeds. In this breed it was found the highest level of effective number of alleles (5,2) and of the observed heterozygosity (0,783). The value of polymorphism index PIC consisted of 0.75 on the average.

The ratio of observed and expected heterozygosity for the most of the studied loci was close to Hardy-Weinberg equilibrium state, the deficiency of heterozygous genotypes was observed only in the loci HTG4 and HTG5. Fixation index was compiled-0108, indicating the absence of inbreeding depression in populations of Kabardin horses.

Tested horses of Trakehner breed (n=49) had a relatively high level of genetic diversity. The number of alleles detected at each locus varied between 4 (HTG7) and 10 (ASB2) and eventually reached 109. The average number of alleles per locus and an effective number of alleles (A_e) was 6,4 and 3,89, respectively (Table 7). The observed heterozygosity (H_o) resulted in 0,719 and ranged between 0,405 (HMS1) and 0,960 (LEX).

This is one of the best sports horse breeds of German origin that has a high coefficient of genetic similarity with Thoroughbred horses (0,829). This is understandable, given the practice of using Thoroughbred and Arabian stallions in the breeding of Trakehner horses. By 1913, most of Trakehner stud stallions were Thoroughbred.

The horses of Hanoverian breed (n=20) showed an average variability for most of the parameters considered (Ae, Ho, MNA). A total of 88 alleles were detected at 17 microsatellite markers among which typical for this warmblood breed were AHT4O (0,275), AHT5N (0,375), ASB2Q (0,450), ASB17N (0,600), HMS3R (0,215), HTG5R (0,375) and LEX3P (0,353). Comparison of Trakehner and Hanoverian horse breeds showed the same range of alleles at loci AHT5, ASB17, ASB23, HTG4, VHL20 and LEX3.

In Table 7 there are shown the summarized the mean number of alleles (MNA), the effective number of alleles (Ne) per loci, the heterozygosity (Ho and He) and the F values of saddle horse breeds bred in Russia. For the microsatellite loci tested across populations, the lowest polymorphism level (3,05) and observed heterozygosity (0,616) were found in Arabian horses.

Table 7 Statistical parameters of saddle horse breeds based on 17 STR loci.

Breed	n	A_e	H_o	H_e	F_{is}	MNA
Akhal-Teke	2024	3,87	0,696	0,697	0,001	7,65
Arabian	420	3,05	0,616	0,641	0,039	6,14
Budenny	33	3,49	0,654	0,688	0,049	5,82
Don*	17	3,390	0,651	0,683	0,047	5,53
Hanoverian	20	3,46	0,711	0,670	-0,061	5,18
Kabardin	303	5,212	0,783	0,718	-0,108	9,71
Tersk	15	3,04	0,628	0,623	-0,008,	4,53
Thorough bred	1945	3,51	0,689	0,688	-0,001	6,00
Trakenen	49	3,89	0,719	0,718	-0,001	6,41

Of the Russian saddle breeds, Kabardin breed showed the higher variability for the most of the considered parameters. The Tersk horse showed low levels of microsatellite variation that may be due to the founder effect (only 2 stallions were used at the initial stage of breeding) in combination with a small population's size.

3.2 Genetic variability of microsatellite loci in the Trotters

Comparative analysis of four Trotter breeds showed that Orlov Trotter breed has the highest alleles variations 17 panel STR loci (a total 132) and mean level of observed heterozygosity (Ho=0,692). The numbers of alleles at loci range from 4 (HTG7) to 13 (ASB17), the most of loci show high level of polymorphism. In Orlov Trotter 12 alleles not found in other Trotters have been typed, including AHT4M, AHT5P, AHT5Q, ASB17H, ASB17P, ASB23M, HMS2Q, HTG4Q, HTG6M, CA425G, CA425L and VHL20J (Table 8). Rare allele AHT5P have not been detected in horses of other breeds except Budenny horses.

Genetic structure of Orlov Trotter is characterized by a high frequency of alleles AHT4O (0,420), AHT5O (0,431), ASB2K (0,314), ASB17R (0,308), ASB23S (0,237), HMS1M (0,578), HMS3P (0,408), HMS6P (0,460), HTG4M (0,597), HTG6O (0,674), HTG10O (0,428) and VHL20M (0,281). The observed heterozygosity (H_o) ranges between 0,546 (HTG6) and 0,834 (ASB17).

Obviously a high level of genetic diversity in Orlov Trotter breed is explained by the fact that it carries the blood of many European horse breeds. In the last century to enhance the playfulness of Orlov Trotter horses were crossed with Thoroughbred and Standardbred, which contributed to the increase in the level of genetic diversity. Interestingly, that the all ten of the best Orlov Trotter sires carry the blood of Thoroughbred and Standardbred horses.

All four Trotting breeds are characterized by the same spectrum of alleles at loci ASB2, HMS6, HMS7, HTG4 and LEX3 and high frequency alleles AHT4O, HMS3P, HTG7O and ASB17R.

The lowest level of polymorphism of microsatellite loci (Ae, Ho and MNV) has been found in Standardbred breed (Table 8 and 9). Obviously this is due to the long focus of breeding American Trotter on actually one quality - fast trot.

Genetic structure of Russian Trotter breed to a large extent is similar to the Standardbred racehorse which for decades was used in its formation.

Comparative analysis of four Trotter breeds at STR loci show that greatest genetic differences and low coefficient of genetic similarity (0,523) are between Orlov and Standardbred trotters. Russian Trotters show higher coefficient of genetic similarity with American Trotters (0,887) than Orlov Trotters (0,631) that strictly confirm their origin. French Trotters have the same high coefficients of genetic similarity (0,847) with the American and Russian Trotters, which form a common cluster.

Table 8 The range of alleles of 17 microsatellite loci in Trotters

Loci	Breed of horses							
	Orlov Trotter n = 1233		Standardbred n = 102		Russian Trotter n = 143		French Trotter n = 61	
	Typical alleles p > 0,05	Rare allele alleles q < 0,05	Typical alleles p > 0,05	Rare allel alleles q < 0,05	Typical alleles p > 0,05	Rare allel alleles q < 0,05	Typical alleles p > 0,05	Rare allel alleles q < 0,05
AHT4	H, J, L, M, O	I, K, P	H, J, O, P	I, K	H, J, O, P	I, K, L	H, I, J, O, P	K
AHT5	J, K, M, N, O	L, P, Q	J, K, M, N, O		J, K, M, N, O		J, K, M, N	O
ASB2	K, M, N, Q	I, O, P, R	K, M, N, O, R	I, P, Q	K, M, N, O, P, Q, R	I	K, M, N,O, P, Q, R	I
ASB17	G, H, I, J, M, N, R	F, K, O, P, Q, S	F,G, M, N, O, R	K, S	F, G, M, N, O, R, S	I, J, K, Q	G, N, O, R, S	F, K, M

38

ASB23	I, J, K, L, S	G, M, T, U	I, J, K, L, G	T, U	I, J, K, L, U	G, S, T	I, J, K, L	T, U
CA425	J, M, N, O	F, G, I K, L	F, J, K, M, N, O		J, K, M, N, O	F, I	J, K, M, N, O	F
HMS1	I, J, M	K, L, N, Q	I, J, K, M	N, Q	I, J, K, M	Q	I, J, K, M	Q
HMS2	H, K, L, R	I, J, M, O, P, Q	K, L, O, P, R	H, J	H, K, L, O, P, R	I, J	J, K, L, O, P, R	H, I, M
HMS3	I, N, O, P, Q	M, R	I, N, P, Q, R	M	I, M, N, P, Q	O, R	I, M, N, P, R	Q
HMS6	K, M, O, P	L	K, M, O, P	L	L, M, O, P	K	K, L, M, O, P	
HMS7	L, M, N, O	J, K	J, L, N	K, O, Q	J, L, N, O	K, M	J, L, M, N	K, O
HTG4	K, L, M,	N, O, P, Q	K, L, M, N, O		K, L, M, N, O	P	K, L, M, N, O	
HTG6	G, J, O,	I, M, P	G, J	I, O, P	G, J, O		G, J, O	
HTG7	K, N, O	M	K, M, N, O		K, M, O, N		K, M, N, O	
HTG10	L, M, O	I, K, P, R, S	I, M, O, R	L, N, P	I, M, O, R	K, P, S	I, K, M, N, O, R	P
LEX3	F, L, M, N, O	H, I, K, P	F, H, L, M, N	I, K, O, P	F, H, L, M, N	I, K, O, P	F, I, K, L, M, P	H, N, O
VHL20	L, M, N, O, P, R	I, J	L, M, N, R	I, Q, P	L, M, N, R	I, O, P	L, M, N, R	I, O, Q

Table 9 Statistical parameters of Trotters based on 17 microsatellite loci.

Breed	n	A_e	H_o	H_e	F_{is}	MNA
French Trotter	61	3,58	0,697	0,695	-0,004	6,41
Orlov Trotter	1233	3,74	0,692	0,706	0,020	7,76
Russian Trotter	143	3,42	0,670	0,679	0,029	6,65
Standardbred	102	3,41	0,649	0,661	0,018	6,24

A_e – effective number of alleles; H_e –expected heterozygosity; H_o – observed heterozygosity
F_{is} – population inbreeding level; MNA- mean number alleles per locus.

Analysis of genetic diversity of Orlov Trotters from different regions shows that subpopulations of Central Russia, Ural and Siberia are characterized by a high degree of genetic similarity and form a common cluster (Fig. 24), indicating the intensive exchange of breeding material throughout the country. However, it were found genetic differences between Orlov Trotters from four leading studs, two of which had formed their interbreed types (Altai stud and Novotomnikovsky stud). Genetic distinctions of stud subpopulations display existence as well as frequencies of alleles, level of polymorphism and heterozygosity. Earlier in the practice of breeding of Orlov Trotter horses it was even used such technique as «geographical heterosis». Genetic differentiation of horses from different studs contributed to maintaining a high level of genetic diversity in this breed.

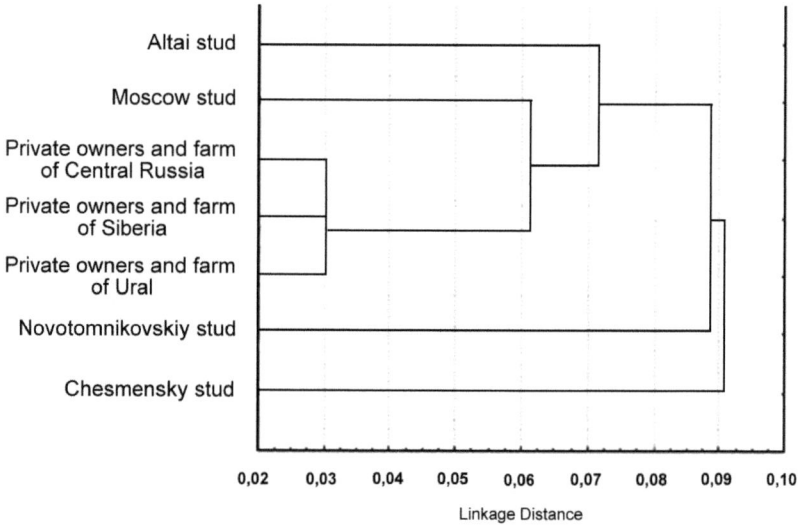

Figure 25 The dendrogram of genetic distances of Orlov Trotters of different lines.

Analysis of allele pool of Orlov Trotter breed at STR loci revealed the presence of a sufficient high level of genetic diversity that can be maintained by improving the linear structure of the breed, the creation of stud types and a wide range of genetic differences between genotypes of trotters of different lines and breeding farms. That provides important additional information about selection processes in the breed that allows you to develop and implement a science-based strategy for its conservation and improvement.

3.3 Genetic structure of domestic Heavy Draft horse breeds

In the middle of the XX century in Russia were created three new breeds of heavy draft horses that have not lost their importance even at the present time. The mainly stallions of Brabant, Ardennas, Percheron, Clydesdales and Suffolk breeds have been introduced from Western Europe since the second half of the XIX century in the view of the industrial development and agricultural intensification. Their offspring resulted from the crosses with indigenous horses, became the foundation for the creation of three different breeds of heavy draught horses dubbed Vladimir Heavy Draft, Russian Heavy Draft and Soviet Heavy Draft. The study of protein polymorphism and blood group systems at horses this heavy draft breeds showed that all three populations are characterized by a high level of genetic diversity and differentiated from each other (Dubrovskaya et al., 1992; Blohina, 2009).

Kalashnikov et all (2014) report the results of the first genetic analysis of Russian heavy draft horses which included the study of the polymorphism of 15 microsatellite loci (AHT4, AHT5, ASB2, ASB17, ASB23, HMS1, HMS2, HMS3, HMS6, HMS7, HTG4, HTG6, HTG7, HTG10 and VHL20). Hair samples were taken from 64 individuals including 15 Vladimir and Soviet Heavy Draft and 34 Russian Heavy Draft horses.

The highest genetic diversity was characteristic of the Russian Heavy Draft horse. In all loci, except HMS6, it was revealed that the maximum number of alleles, a total 98. In horses of draft breeds were identifi 58 common alleles, and only at the HTG7 locus was determined by the same variants of alleles. Soviet and Russian heavy breeds was characterized by high frequency of alleles HTG6O and HTG4L. The lowest allele variability at 15 STR loci (MNA=4,27) was found in Soviet Heavy Draft (Table 10).

The effective number of alleles per locus ranged from 3,15 (Soviet Heavy Draft) to 3,80 (Soviet Heavy Draft). The Vladimir Heavy Draft horses had higher genetic diversity compared to the Soviet Heavy Draft (Ae=3,44).

Of the Russian heavy draft breeds, Russian Heavy Draft shows the higher variability for most of the considered parameters, while the Soviet Heavy Draft showed the lowest variability for microsatellite loci. Perhaps this is due to the limited size of the population of the Soviet Heavy Draft horses and the inevitable flood of inbreeding and a small number of tested horses.

Table 10 Genetic variability for the microsatellite loci in Russian heavy draft breeds

Loci	Number of alleles			Effective number of alleles			Observed heterozygosity		
	VHD	RHD	SHD	VHD	RHD	SHD	VHD	RHD	SHD
AHT4	6	8	4	2,49	4,94	3,57	0,533	0,794	0,800
AHT5	7	7	5	5,29	4,84	3,31	0,867	0,735	0,667
ASB17	8	8	7	5,23	4,77	5,17	0,867	0,794	0,800
ASB2	6	8	5	4,25	4,67	3,98	1.000	0.727	0,933
ASB23	6	8	5	4,05	4,15	4.55	0,800	0,367	0,733
HMS1	4	5	3	1,42	2,49	2,76	0.267	0,618	0,733
HMS2	5	7	4	3,95	3,37	2,57	0,667	0,767	0,600
HMS3	6	7	3	3,44	3,31	2,07	0,733	0,706	0,333
HMS6	6	4	3	3,36	3,35	2,94	0,667	0,500	0,733
HMS7	5	8	5	3,98	4,79	3,46	0,533	0,824	0,667
HTG10	6	7	5	2,49	3,03	3,41	0,733	0,706	0,800
HTG4	4	5	3	2,94	2.73	1,85	0,533	0,706	0,467
HTG6	4	4	1	3,24	1,36	1.00	0,600	0,294	0.000
HTG7	5	4	4	2,03	3,27	3,28	0,467	0,706	0,800
VHL20	6	8	7	3,46	5,94	3.31	0,667	0,823	0,733
Mean	5,60	6,53	4,27	3,44	3,80	3,15	0,662	0,673	0,653

* VHD - Vladimir Heavy Draft, RHD- Russian Heavy Draft, SHD - Soviet Heavy

Negative F_{is} value (-0,037) were found only in Soviet Heavy Draft breed. The Vladimir and Russian Heavy Draft had positive Fis value, which was in accordance with the discrepancy between the gene diversity and observed heterozygosity in these breeds.

Cluster analysis showed a stronger connection between Russian and Soviet Heavy Draft and more significant genetic differences between them and the Vladimir Draft horse.

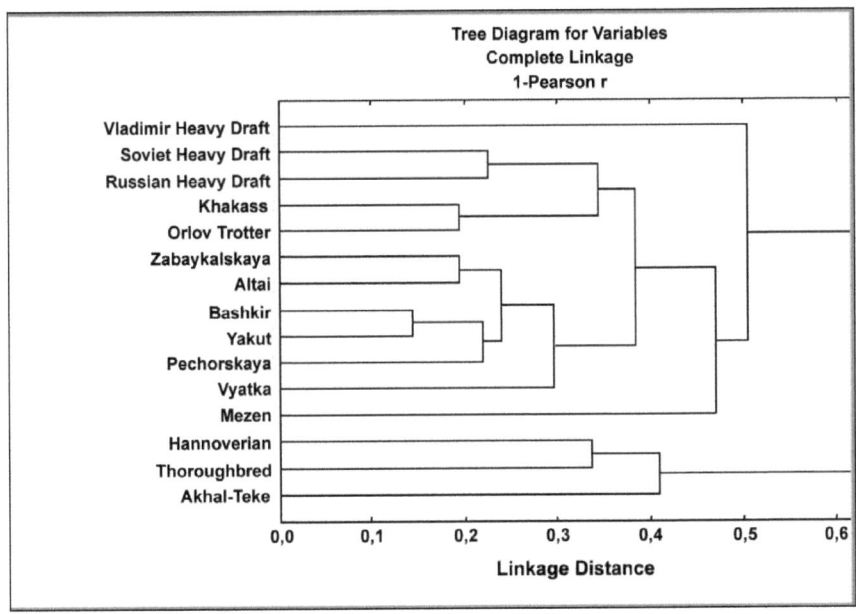

Fig. 26 The dendrogram of genetic distances for heavy draft and another domestic horse breeds (Kalashnikov et al., 2014).-(Kon. 2014, 4; p.8)

On the dendrogram (Fig. 26) Russian and Soviet Heavy Draft breeds form separate subclusters, which can be differentiated from each other. The Vladimir Heavy Draft is genetically unique and markedly different from all other domestic horse breeds. Sequencing fragments of non-coding region of D-loop of mitochondrial DNA in mares of Vladimir Heavy Draft breed revealed 19 haplotypes belonging to 10 groups B, E, G, Y, I, J, L, M, P and Q (Sorokin, 2014). Sequence alignment was performed by using a reference equine mtDNA sequence (GeneBank X79547). Seven of the 19 female families (36.8%) represented the haplogroup B, the three families - haplogroup L and the two families – haplogroup Q. It should be noted that haplogroup L is most prevalent in the horses of European heavy draft breeds.

A comprehensive analysis of mitochondrial DNA from Vladimir Heavy Draft horse mtDNA haplotypes indicates multiple maternal origins of this indigenous breed and its phylogenetic relationships with other European breeds (Fig. 27).

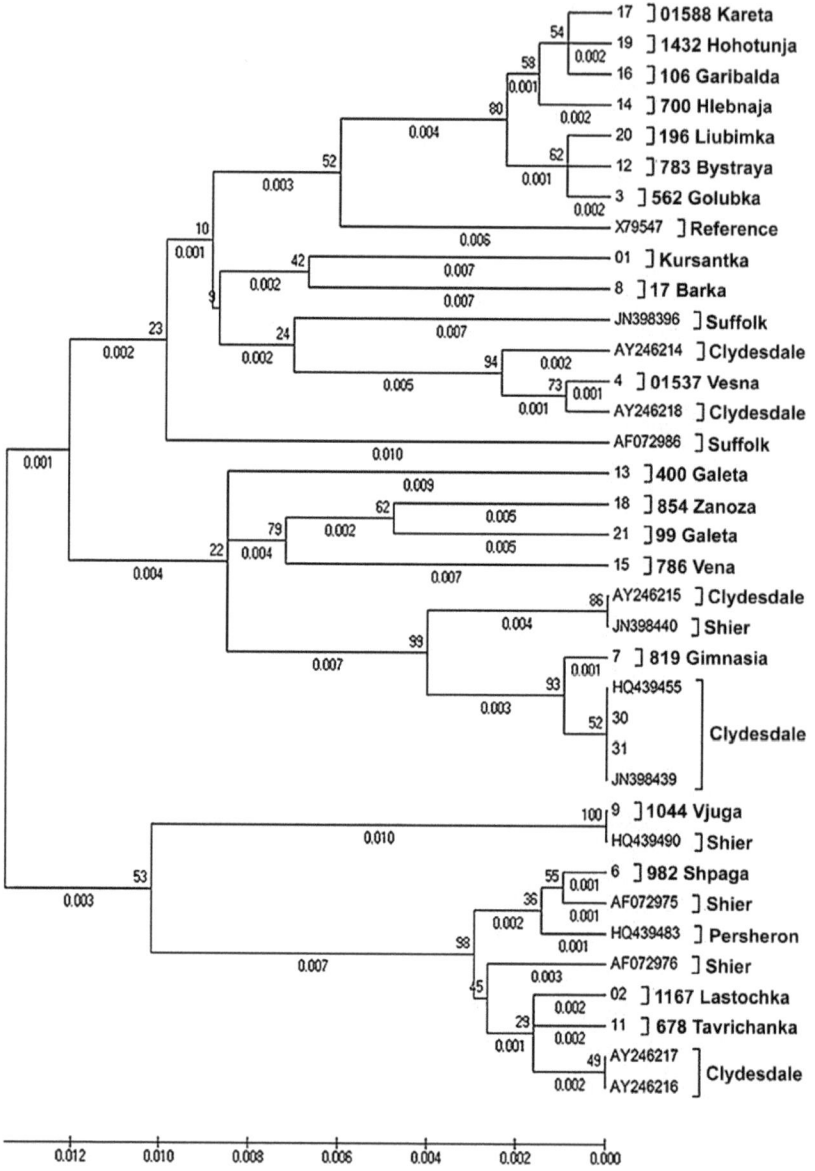

Fig. 27 Phylogenetic relationships in the families of the Vladimir Heavy Draft horses with other European breeds on the basis of haplogroups of mtDNA (Sorokin, 2014).

3.4 Genetic characterization of native horse breeds

The study of polymorphism of STR loci in 10 Russian native horse breeds revealed high genetic diversity among all inspected populations. For comparison Shetland pony (n=32) and the Mongolian horse (n=14) as representatives of the typical native breeds of European and Asian origin were included.

The analysis results for 12 native horse breeds are shown in Table 11 and Table 12.

The number of alleles at each locus varied from 2 (HTG6 in Pechora horse) to 14 (ASB17 in Altai horse) and on average amounted to 6.82 per locus. A total of 184 alleles were detected across the 17 STR loci analyzed in 395 horses of 12 native breeds. The observed number of alleles per population ranged between 141 (Bashkir breed) and 93 (Buryat breed). The number of private alleles per population ranged between zero (Buryat, Khakas, Tuva and Zabaikalskaya breeds) and four (Bashkir breed). The private alleles were registered in 8 native breeds including Bashkir (4), Mezen (3), Altai (2), Yakut (1) Pechora (1), Vyatka (1), Shetland pony (1) and Mongolian horse (2). Most of the private alleles were in very low frequencies below 5%.

In addition to the standardized equine DNA typing allele's nomenclature of 17 equine-specific STR loci there were detected some more alleles, including AHT5P, ASB2H, ASB17U, ASB17Y, ASB23M, ASB23N, CA425E, HMS1O, HMS1R, HMS2D, HMS2G, HMS2N, LEX3R and LEX3S in local horse breeds. There were significant differences in the total number of allele variants, effective number of allele (Ae) and number of allele per loci (MNA) among breeds (Table 12). Population diversity varied in inspected breeds by Ae from 3.21 (Buryat horse) to 4.60 (Altai horse).

heterozygosity (Ho) values among horse breeds ranged from 0,691 (Vyatka) to 0,802 (Mongolian). Only 5 from 12 breeds were in Hardy-Weinberg equilibrium (HWE) and had negative Fis values. The reason of positive Fis values for some observed breeds may be due to small number of tested animal.

45

Table 11 The range of alleles of 17 microsatellite loci in Trotter breeds

Loci	Breed of horses							
	Altai n = 39		Bashkir n=100		Mezen n = 43		Yakut n = 42	
	Typical alleles p > 0,05	Rare alleles q < 0,05	Typical alleles p > 0,05	Rare alleles q < 0,05	Typical alleles p > 0,05	Rare alleles q < 0,05	Typical alleles p > 0,05	Rare alleles q < 0,05
AHT4	H, J, K, O	I, L, N	H, J, M, O	I, K, L, N	H, I, J, K, O	N	H, I, J, O	K, L, M, N, P
AHT5	J, K, L, M, N, O		J, K, L, M, N, O		J, K, N, O	**H**, I, L, M	J, K, L, N, O	I, M
ASB2	B, **C**, K, M, N, O, Q	I, P, R	I, K, M, N, O	**H**, P	K, M, N, O, Q	B, I, P	I, K, M, N, O	L, Q, R
ASB17	G, H, I, M, N, R, **W***	I, K, O, P, Q, S, **V***	I, K, M, N, O, Q, R	**D**, F, G, J, L, P, S, T, **U***	F, H, K, N, O, P, R	L, S, **Y***	I, N, P, Q, R, S	F, H, K, L
ASB23	I, J, K, L, S, U	M, R, T	I, J, K, L, **P***, S, U	G, Q	I, J, K, L, M, S, U	G	J, K, R, S, T, U	L
CA425	F, G, J, M, N, O	I, L	**E***, G, I, J, L, M, N, O	F	J, K, L, M, N		F, G, I, J, M, N	O
HMS1	J, L, M, N	I, **O**	I, J, K, M	L, N	I, J, L, M, N	Q	J, M	I, K, L, N, Q
HMS2	H, I, K, L	M,O, R	H, I, J, K, L	**D***,**G**, O, P, R	H, I, J, K, L	O, R	H, I, K, L, R	O, P
HMS3	I, M, N, O, P, Q	R	I, M, N, P, Q, R	**S**	I, M, P, Q, R	N, O	I, M, O, P, Q, R	I, **K***
HMS6	L, M, O, P	K, N	K, L, M, O, P	Q	K, L, M, N, O, P	**J***	K, M, O, P	L, Q
HMS7	J, K, L, N, O	M, Q	L, M, N, O	K, Q	J, L, M, N, O		L, N, O	J, M, Q
HTG4	K, L, M, O	N, P, Q	L, M, P	K, N, O	K, L, M, O	P, Q	K, L, M, O	
HTG6	G, J, O,	I, M, P	G, J, M, O, P	I, N	G, J, O		G, J, M, O	
HTG7	K, M, N, O		K, M, N, O	L	K, M, N, O		K, M, N, O	L
HTG10	I, K, L, M, O, R, S	P, Q	I, K, L,M, N, O, P, R	T	I, K, M, O, R, S	L, N, T	K, L, O, Q, R	I, M, P
LEX3	F, H, K, L, M, N	I, J, P	H, K, L, M, N,	F, I, O	M, O, P	K, **R, S***	F, I, L, M, N	O, P, Q
VHL20	I, L, M, N, Q, P, R	O	I, J, M, N, P, Q, R	L, O	I, J, M, N, O, R	P, Q	I, M, N, O, P, Q, R	J, L

*Private alleles

46

Table 12 Statistical parameters of 15 horse breeds based on 17 microsatellite loci.

Breed	n	A_e	H_o	H_e	F_{is}	MNA
Altai	39	4,60	0,723	0,750	0,036	7,47
Bashkir	100	4,44	0,755	0,750	-0,006	8,29
Buryat	13	3,21	0,694	0,697	0,004	5,47
Khakas	18	4,05	0,726	0,727	0,001	6,12
Mezen	43	3,81	0,693	0,709	0,023	6,59
Pechora	12	4,25	0,726	0,738	0,016	5,81
Tuva	35	4,20	0,748	0,742	-0,008	6,65
Vyatka	16	3,72	0,691	0,709	0,025	6,00
Yakut	42	4,27	0,734	0,732	-0,003	7,00
Zabaikalskaya	31	4,01	0,729	0,732	0,004	6,82
Shetland	32	3,59	0,701	0,685	-0,023	6,19
Mongolian horse	14	5,16	0,802	0,791	-0,014	7,65

Ae – effective number of alleles; He – expected heterozygosity; Ho – observed heterozygosity; Fis – population inbreeding level; MNA – mean number alleles per locus.

The allelic variations of microsatellite loci were insignificantly higher for the inspected native breeds than for the cultural breeds. Russian native breeds showed more high level of polymorphism of the X-chromosomal locus LEX3. There were determinate all 12 alleles at that locus, including LEX3J (Altai and Mongolian), LEX3P (Yakut and Mongolian), LEX3Q (Yakut), LEX3R (Mezen and Mongolian) and LEX3S (Mezen), which are absence in cultural horse breeds. That indicates of wider evolution platform in ancient populations of horses.

Among Russian native horse breeds the Bashkir horses showed the higher variability for the most of the considered parameters (except for Ae, where the Altai horse had the highest value). Testing Bashkir horses on blood groups and proteins also showed scene high level of genetic diversity (Dubrovskaya et al., 1992; Khrabrova, Zaitcev 2008).

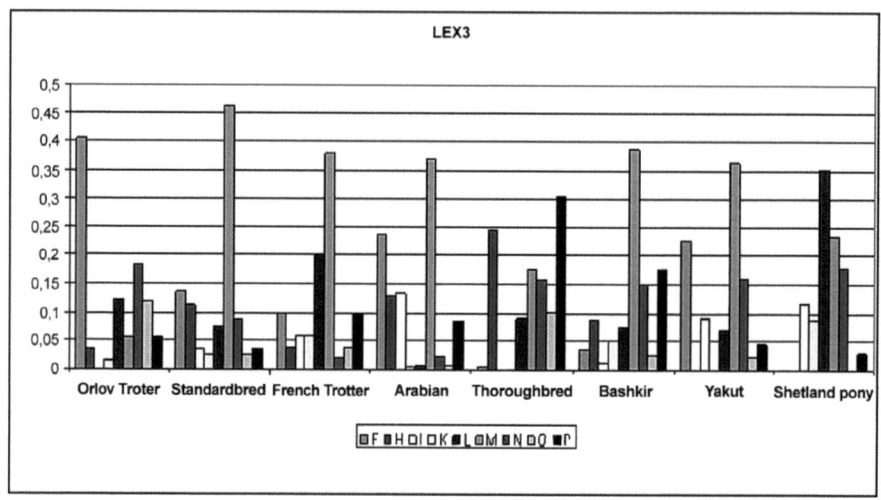

Figure 28 Allele variability at locus LEX3 in cultural and native horse breeds.

Genetic structure of Bashkir horse is characterized by a high frequency of alleles AHT5K (0,313), ASB2KQ (0,294), HMS1M (0,578), HMS3P (0,408), HTG10O (0,437), CA425M (0,337), LEX3M (0,387) and VHL20P (0,162). Obviously a great variety of alleles of microsatellite loci from Bashkir horses resulted due to both large number of the population and the influence of cultural breeds. This is confirmed by the relatively high coefficients of genetic similarity between Bashkir horse and Orlov Trotter (0,858) and also Russian Heavy Draft horse (0,806) as well. Bashkir horse breed showed the highest genetic similarities with Yakut horse (0,891) and maximum differences with the Thoroughbred (0,400).

Genetic structure of the endangered Altai horse also was characterized by high allelic richness (Ae = 4,6) and private alleles at loci ASB17W (0,128) and ASB17V (0,013). These rare alleles were determined only in Altai from Ulagan population and were absent in horses from Kosh-Agach region located near the broad with China. Altai horses are characterized by comparatively high concentration of alleles ASB2C

(0,056), HMS3Q (0,158), HMS3L (0,412), HTG4M (0,667), VHL20M (0,346) and LEX3F (0,272). Altai horse breed showed the highest genetic similarities with the Bashkir horse (0,880) and maximum differences with the Don horse (0,489).

Vera Warmuth et al (2012) also determined two private alleles and high level of genetic diversity at STR loci in Altai horse (n=40). The analyses of non-bred horse populations in Asia and parts of Eastern Europe revealed a pattern of isolation by distance and significant decline in genetic diversity (expected heterozygosity and allelic richness) from east to west, consistent with a westward expansion of horses out of East Asia.

Conducted by Keyser-Tracqui et al (2005) analysis mtDNA of 13 ancient horses recovered from a frozen tomb near the village of Berel (Altai, east Kazakhstan) determined 8 haplotypes, found in horses in Europe and the Middle East. An ancient Scythian burial located near the border of Russia and China and the area of the Altai horses and therefore their study is of undoubted interest for the analysis of the origin of horse breeds of Eurasia.

In comparison with other North European breeds Mezen horse show high level of genetic diversity for this small and geographically isolated population. Genetic structure of Mezen horse breed is characterized by high concentration of alleles AHT4I (0,209), HMS3K (0,412), HMS7L (0,686), HTG6O (0,779), ASB23S (0,325) and LEX3M (0,666). In addition allele spectrum of this breed include the rare variants ASB17Y (0,047), HMS6J (0,012), LEX3R (0,048) and LEX3S (0,048). The origin of Mezen horse is still being discused, genetic analyses suggests of unique this horse population (Fst=0,063). This forest breed showed the highest genetic similarities with the Bashkir (0,841) and Pechora (0,811) horses. The most numerous in Russia and obviously in the world the Yakut horse (n=170.000) occupies a vast area in Central Siberia and can be considered an ancient breed of Northern Asia. Genetic structure of Yakut breed is characterized by a high level of genetic diversity all 17 STR loci (Ae=4,27, Ho=0,734). For the Yakut horses are typical relatively high frequency of alleles HTG4M (0,637), VHL20Q (0,205), LEX3M (0,364) and also the presence of private allele HMS3K (0,023). Rare allele LEX3Q was found only in Yakut and Mongolian horses.

It is interesting that Yakut horse breed showed higher genetic similarities with the European native horse breeds such as Bashkir (0,891) and Pechora (0,837) than Mongolian (0,835). Yakut horse has a medium level of genetic differentiation compared with other breeds (Fst = 0,033) and is most different from the Don (0,539) and Tersk (0,538) horses.

The studied horse populations differed in their genetic structure and degree of differentiation. Nei's genetic distances were in interval 0,075 - 0,690. The dendrogram of Figure 29 shows the genetic relationship among horse breeds from restricted maximum likelihood analysis of the gene frequency data at 17 microsatellite loci. The analysis showed that all Russian native horse breeds from Eastern Europe and Asia form overall cluster with Orlov Trotter and Russian Heavy Draught breeds.

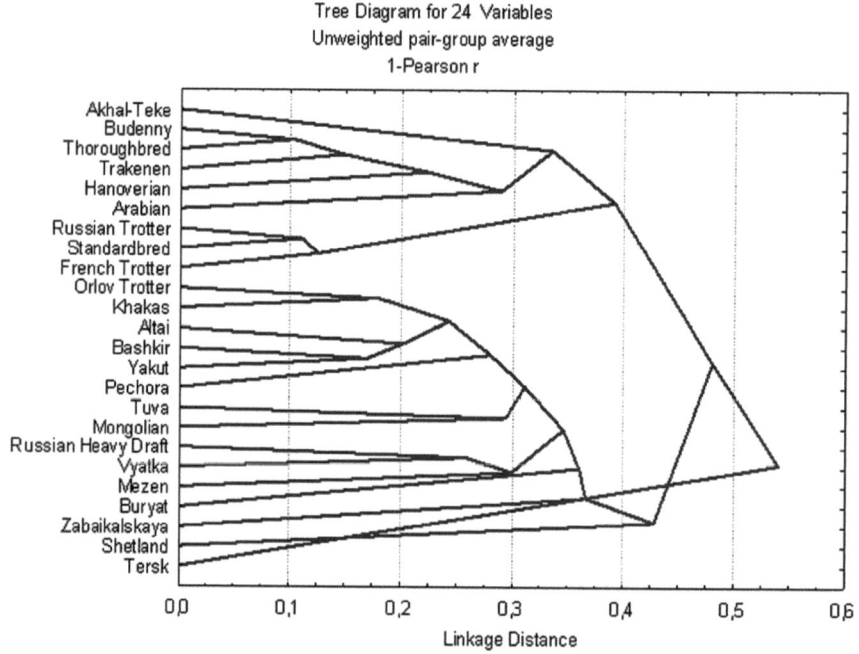

Fig. 29 Dendrogram of Nei's genetic distances between Russian horse breeds at 17 STR loci.

All riding horse breeds, except Tersk breed, enter into a sub-cluster in which the Akhal-Tekes formed a separate line. Warmblood Budenny breed has demonstrated its close relationship with the Thoroughbred resulted from systemic crossbreeding with Thoroughbred stallions.

Three trotter breeds including the American, Russian and French Trotters, formed a common subcluster, which is consistent with the history and breeding

50

system of these populations. Coldblooded Orlov Trotters and to a lesser extent Russian Heavy Draft horses were used to improve of the many native breeds for a long time, which of course had an impact on the genetic structure of many native populations.

The Mongolian horse is in the center of subclusters of horse breeds of Russian origin. In the middle ages, this horse together with the army of Genghis Khan penetrated into almost in all the States of Eurasia and has undoubtedly influenced the process of rock formations for this vast territory. Cluster analysis showed that the local breeds of horses differ markedly from Shetland pony, which indicates the existence of differences between local populations of Western and Eastern Europe.

CONCLUSION

A genetic analysis of 27 stud and native horse breeds that are bred in Russia demonstrated high diversity at 17 panel microsatellite loci in many populations. A total of 184 alleles were detected including some additional variants AHT5P, ASB2H, ASB17U, ASB23M, ASB23N, CA425E, HMS1O, HMS1R, HMS2D, HMS2G, HMS2N, LEX3R and LEX3S. The mean number of alleles per locus among the horse breeds ranged from 4,53 (Tersk) to 9,71 (Kabardin) and the observed heterozygosity ranged from 0,616 (Arabian) to 0,755 (Bashkir). The private alleles were registered in some breeds including Altai, Bashkir, Kabardian, Mezen, Pechora, Vyatka and Yakut. Native breeds of horses had a relatively high level of interbreed differentiation, indicating the need of studying multiple subpopulations to complete information about the breed. The inbreeding coefficient had negative or low value for 67% populations.

The studied horse breeds differed in their genetic structure and degree of differentiation (Fst= 0,009-0,177). The analysis showed that the most of Russian horse breeds, including coldblooded Orlov Trotter and Russian Heavy Draft and all native populations form overall subcluster. Obviously, in the middle of the last Millennium the Mongolian horse had a definite impact on many breeds and populations of horses of Russia and of the whole Eurasia. All native populations of horses, including endangered Altai, Mezen, Pechora and Vyatka breeds, revealed rather high resources of genetic variability permitting successful implementation of conservation programs. It is important to concern the genetic diversity of endangered stud and native horse breeds on the basis of effective management, especially in respect to small population sizes.

BIBLIOGRAPHY

Animal Genetic Resources for Food and Agriculture. The State Of The World's (2007). Commission On Genetic Resources for Food and Agriculture Organization of the Unitad Nations. Rome.

Arabian horse (2003). Centropolitgraf, Moscow.

Barmintsev Yu. N., Kozhevnicov Ye. V. (1983). Horse breeding in the USSR. Kolos, Moscow.

Bjornstad G., Roed K.H. (2002). Evaluation of factors affecting individual assignment precision using microsatellite data from horse breeds and simulated breed crosses. Animal Genetics 33 (4), 933-944.

Binns M., Swinburne J.E., Breen M. (2000). Molecular Genetics of the Horse. In: The Geenetics of the Horse. CABI Publishing, 109-121.

Bjornstad G., Roed K.H. (2001). Breed demarcation and potential for breed allocation by microsatellite markers. Animal Genetics. 32, 59-65.

Bjornstad G., Nilsen N.O., Roed K.H (2003). Genetic relationship between Mongolian and Norwegian horses. Animal Genetics. 34 (1), 55-58.

Blohina N.V. (2009). The genetic characteristics of the populations of Heavy Draft horses breeds. Konevodstvo i Konny Sport 5, 14-16.

Bowling A.T., Ruvinski A. (2000). The Genetics of the Horse CABI Publishing. Wallingford.

Breen M., G Lindgren G., Binns G. et al (1997). Genetical and physical assignments of equine microsatellits - first integration of anchored markers in horse genome mapping. In: Mammalian Genome 8, 267-273.

Campana M.G., Stock F., Barrett E., Benecke N., Barker G.W.W., Seetah K., Bower M.A. (2011). Genetic stability in the Icelandic Horse Breed. Animal Genetics 43, 447-449.

Canon J., Checa M.L., Carleos C., Vega-Pla J.L., Vallejo M., Dunner S. (2000). The genetic structure of Spanish Celtic horse breeds inferred from microsatrllite data.

Animal Genetics 31 (1), 39-48.

Cothran E.G., Luis C. (2005). Genetics distance as a tool in the concervation of rare horse breeds. In: Conservation Genetics of Endangered Horse Breeds (Ed. By

I. Bodo, L. Alderson, B. Langlois). P. 55-71. Wageingen Academic Publishers, The Netherlands.

Cunningham E.P. (2005). Molecular methods and equine genetic diversity. In: Conservation genetics of endangered horse breeds. EAAP publication No 116, Bled, Slovenia, 15-24.

Dergunova M.M., Kolomeets J.J., Khrabrova L.A. (2012). Molecular-genetic features of Khakassian horse Konevodstvo i Konny Sport 6, .8-9.

Dimsoski P. (2003). Development of a 17-plex Microsatellite Polymerase Chain Reaction Peaction Kit for Genjtyping Horses. Croatian Med. J. 44(3), 332-335.

Dubrovskaya R. M., Starodumov I.M., Bannikova L.V. (1992). Genetic differentiation of horses breeds at polymorphic loci of blood proteins. Genetics 28 (4), 152-165.

Duduev A. S., Houdov A. D., Kokov Z. A., Umshokov H. K., Jekamukov M. H., Zaitsev A. M., Zaitseva M. A., Gavrilicheva I. S., Kalinkova L. V. and Reissman

M. (2014). Genetic structure of the Kabardin breed of horses by the loci of the microsatellite DNA and methods for identifying populations. Konevodstvo i Konny Sport 6, .18-19.

Goldstein D.B., Linares A.R., Cavalli-Sforza L.L. (1995). An evelution of genetic distances for use with microsatellite loci. Genetics 139, 463-471.

Goldstcin D.B., Schlottcrer C. (1999). Microsatellits: Evolution and application. N.Y.: Oxford Univ. press.

Guerin G. Bertand M., Billoud B., Meriaux J.C. (1993). A genetical analisis of variable number tandem repeat (VNTR) polymorphism in the horse. In: Genetics Selection and Evolution 25, 435-445.

Guerin G. Balley E., D. Bernoco D. et al. (2003). The second generation of the Internetional Equine Gene Mapping Workshop half-sibling lincage map. Animal Genetics 34, 161-168.

Edwards E.H. (1991).The ultimate horse. Dorling Kindersley Limited, London.

Ellegen H., Johansson M., Sandberg K., Andersson L.(1994).Cloning of highly polymorphic microsatellites of the horse. Animal Genetics 23, 133-142.

Iwanczyk E., Juras R., Cholewinski G., Cothran E.G. (2006). Genetic structure and phylogenetic relationships of the Polish Heavy Horse. J. Appl. Genet. 47(4), 353-359.

Jansen, T., Foster P,. Levine M.A., Oelke H., Hurles M., Renfrew C., Webewr J., Olek

K.1 (2002). Mitochondrial DNA and the origin of the domestic horse. Proc. of the National Acad. of Sci. USA 99, Issue 16, 10905-10910.

Juras R. (2005). Genetic analysis of Lithuanian native horse. Summary Doct. Diss. Biomed. Sci., zootechny (13 B), Kaunas.

Kalashnikov V.V., Khrabrova L.A., Zaitcev A.M., Zaitceva M. A., Kalinkova L.V. (2011). Polymorphism of microsatellite DNA in horses of stud and local breeds.

Sel'skokhozyaistvennaya biologiya 2, 41-45.

Kalashnikov V.V., Khrabrova L.A., Zaitcev A.M. (2013). Applied genetics for horse breeding. J. Farm Animals 2, 60-62.

Kalashnikov V.V., Khrabrova L.A., Zaitceva.M.A., Gavrilicheva I.S., Kalinkova L.V., Kalinkina G.V., Makhmutova O.N. (2014). Using microsatellite loci of DNA for assessing genetic diversity of Orlov trotting breed of horses. Bulletin of the Russian Academy of Agricultural Science 2, 30-33.

Kalashnikov V.V., Zaitcev A.M., Kalinkova L.V. (2014). Genetic structure of three domectic Heavy Draft horse breeds and their genetic differentiation at DNA microsatellite loci. Konevodstvo i Konny Sport 4, 6-8.

Kashi Y., Soller M. (1999). Functional roles of microsatellites and minisatellites. In: Microsatellites. Evalution and application. N.Y.: Oxford Univ. press Inc., 10-23.

Keyser-Tracqui C., Blandin-Frappin P., Francford H.-P., Ricaut F-X., Lepetz S., Crubezy E., Samashev Z., Ludes B. (2005). Mitochondrial DNA analysis of horses recovered from a frozen tomb (Berel site, Kazakhstan, 3rd Century BC). Animal Genetics 36(3), 203-215.

Khrabrova L.A., Zaitcev A.M. (2002). The genetic structure of Vyatka horse populations. Proceeding of the XXVIII Inter. Conf. on Animal Genetics. Gottingen - Germany, 109.

Khrabrova, L.A. (2008). Monitoring of the genetic structure of breeds in horse breeding. J. Russian Agricultural Science 34 (4), 261-263.

Khrabrova L.A., Zaitcev A.M. (2008). Features of allele spectrum in local horse breeds. Konevodstvo i Konny Sport 3, 19-21.

Khrabrova L.A., Kalinkova L.V., Zaitseva M.A. (2008). Genetic differentiation of thoroughbred horse breeds in microsatellite loci. Sel'skokhozyaistvennaya biologiya 2, 31-34.

Khrabrova L.A., Kalinkova L.V., Zaitseva M.A., Zaitsev A M. (2008).Polymorphism of 17 microsatellite loci in Akhal-Teke, Arabian and Thoroughbred horses in Russia. Proceeding of the. XXXI Inter. Conf. ISAG. Amsterdam, 2043.

Khrabrova L.A., Kurnyavko N.U., Sotnikova S.A. (2012). Characteristics of polymorphism of microsatellite loci in horses of Budenny breed. Konevodstvo i Konny Sport 3, 6-8.

Khrabrova L.A., Zaitseva M.A. (2013). Polymorphism evaluation of microsatellite markers in native Russian horse breeds. Proc. of the 64 Annual Meeting of European Federation of Animal Science. Nantes, France. 2013, 322.

Khrabrova L., Zaitsev A., Zaitseva M., Kalincova L., Gavrilicheva I. (2014). Characterization of genetic horse breeding resources in Russia using STR markers. J. Animal Husbandry 62, 14-20.

Klukowska-Rotzler J., Jost U., Schelling C., Dolf G., Clowdhary B.P, Leeb T., Gerber V. (2006). Characterization and RH mapping of six gene-associated equine microsatellite markers. Animal Genetics 37(3), 305-307.

Koban E., Denizci M., Aslan O., Aktoprakligil D., Aksu S., Bower M., Balcioglu B.K., Ozdemir Bahadir A., Bilgin, Erdag B., Bagis II., Arat S. (2011). High microsatellite and mitochondrial diversity in Anatolian native horse breeds shows Anatilia as a genetic conduit between Europe and Asia. Animal Genetics 43 (3), 401-409.

Kozlov S. A., A. Parfenov V.A. (2012). Horse breeding. KolosS, Moscow.

Kruger K., Stranzinger G., Rieder S. (2002). A full genom scan panel of horse (Equus caballus) microsatellite markers applied to different equid species. Proc. XXVIII International Conf. on Animal Genetics.Gottingen- Germany. 113.

Ling Y.H., Ma Y.H., Guan W.J., Cheng Y.J., Wang Y.P., Han J.L., Mang L., Zhao Q. J., He X.H., Pu y.B., Fu B.L. (2011). Evaluation of the genetic diversity and population structure of Chinese indigenous horse breeds using 27 microsatellite loci. Animal Genetics 42(1), 56-63.

Lukashov V.V. (2009). Molecular evolution and phylogenetic analysis. Benom, Laboratory of Knowledge. Moscow.

Luis C., Juras R., Oom M.M., Cothran E.G. (2007). Genetic diversity and relationships of Portuguese and other horse breeds based on protein and microsatellite loci variation Animal Genetics 38(1), 20-27.

McGahern A., Bower M.A.M., Edwards C.J., Brophy P.O., Sulimova G., Zakharov I., Vizuete-Forster M., Levine M., Li S., MacHugh D.E., Hill E.W. (2006). Evidence for biogeographic patternining of mitochondrial DNA in Eastern horse populations.

Animal Genetics 37(5), 494-497.

Mickelson J.R., Petersen J.L., Mccue M.E. (2012). Book of Abstracts of the 63rd Annual Meeting of the European Federation of Animal Science. No 18. Bratislava, Slovakia. Wageningen Academic Publishers, 326.

Moodley Y. Baumgarten I., Harley E.H. (2006). Horse microsatellites and their ability to comparative equid genetics. Animal Genetics 37(3), 258-261.

Nei M. (1987). Molecular evolutionary genetics. N. Y.: Columbia Univ. press.

Nei M., Takezaki N. (1996). Reconstruction of phylogenetic trees from microsatellite (STR) loci.. Proc. XXV Inter. Conf. of Animal Genetics. Tours-France, 2-3.

Ollivier L., Chevalet C., Foulley J.L. (2005). The use of markers for characterizing genetic resources. In: Conservation genetics of endangered horse breeds. EAAP publication No116. Bled, Slovenia, 25-35.

Nicholas, F.W. (2003). Introduction to veterinary genetics. 2nd ed. Dlacrwell Publishing Ltd.

Perez-Gutierrez L.M., De la Pena A., Arana P. (2008). Genetic analysis of the Hispano-Breton heavy horse. Animal Genetics 39(5), 506-514.

Saastamoinen M.T., Mäenpää M. (2005). Rera horse breeds in Northern Europe. In: Conservation genetics of endangered horse breeds. EAAP publication No 116, Bled, Slovenia, 129-136.

Seyedabadi H, Amirinia C., Banabazi M.H., Emrani H. (2006). Parentage verification of Iranian Caspian horse using microsatellite markers. Iranian J. of Biotechnology 4(4), 260-264.

Rozdestvenskaja G.A., Kalinkina G.V., Orlova Y. Kalinkova L.V., Kreshihina V.V. (2012). Features of the origin and dynamics of female lines (families) in Orlov Trotter breed. Konevodstvo i Konny Sport 1, 8-10.

Ryabova T.N., Khrabrova L.A., Ustyantseva A.V., Morozov R.O. (2012) Genetic diversity evaluation of horse populations of Akhal-Teke breed on DNA markers. Konevodstvo i Konny Sport 5, 6-8.

Sorokin S.I. (2014). Selection and genetic methods to impove Vladimir Draft horse breed in condition of a limited gene pool. Diss. Agricultur. Sci. Divovo, Russia.

Sponenberg, D.P. (2000). Genetic resourses and thear conservation. In: A.T. Bowling,

A. Ruvinski Genetics of the horse. CABI Publishing, Wallingford, 387- 410.

Sponenberg, D.P. (2003). Equine color genetics. 2nd ed. Blackwell Publishing.

Szontagh A., Ban B., Bodo I., Cothran E.G., Hecker W., Jozsa Cs., Major A. (2005). In: Conservation genetics of endangered horse breeds. EAAP publication No 116, Bled, Slovenia, 123-128.

Tautz D. (1989). Hypervariability of simple sequences as a general source for polymorfic DNA markers. Nucl. Acids Res. 17, 6463-6471.

Thirstrup J.P., Petroldi C., Loeschcke V. (2008). Genetical analysis, breed assignment and conservation priorities of three native Danish horse breeds. Animal Genetics 39(5), 496-505.

Van de Goor L.H.P., Panneman H., van Haeringen W.A. (2010). A proposal for standardization in forensic equine DNA typing: allele nomenclature for cquine- specific STR loci. J. Animal Genetics 41, 122-127.

Van de Goor L.H.P., van Haeringen W.A., Lenstra J.A. (2011). Population studies of 17 equine STR for forensic and phylogenetic analysis. Animal Genetics 42, 627-633.

Vitt V. O. (1952). From history of Russian horse breeding. Moscow, Russia. Agro State Publishing.

Warmuth V., Manica A. Eriksson A., Barker G., Bower M. (2012). Autosomal genetic diversity in non-breed horses from eastern Eurasia provides insights into historical population movements. Animal Genetics 44(1), 53-61.

Weller J.I., Seroussi E., Ron M. (2006). Estimation of the number of genetic markers required for individual animal identification accounting for genotyping errors. Animal Genetics 37(4), 387-389.

Wright J.M., Bentzen P. (1994). Microsatellites: Genetics markers for the future. Rev. Fish. Biol. Fish. 4, 384-388.

Zaitseva M.A., Khrabrova L.A., Kalinkova L.V. (2010). Intrabreed diversity on 17 loci of microsatellite DNA in Arabian horses of different lines. Konevodstvo i Konny Sport 1, 19-21.

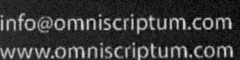